Accepted Science & Paradigms Which Are Likely Wrong

Copyright Page

This book is copyrighted for 2022

Title: Accepted Science & Paradigms Which Are Likely Wrong

The Crazy and Out of the Box Series Book 12

By Martin K. Ettington

All Rights Reserved USA 2022

ISBN: 9798847849401

Printed in the United States of America

Accepted Science & Paradigms Which Are Likely Wrong

Accepted Science Which is Wrong or Limited

As an engineer and student of science I try to keep up with publications of the latest discoveries and paradigm shifts in science around the world.

I believe that there are several big problems with science today. First is that much science is biased which might be because some scientists want to keep their point of view prominent and the others because many focuses of science are supported by biased funding. This will be discussed in this book.

The other issue is that many scientists limit their thinking to only support age old paradigms and refuse to accept changes to their theories or throw out evidence because it doesn't conform their current beliefs.

In this book we will cover many issues such as the belief among many cosmologists that the Big Bang never happened.

We will also look at ideas we thought settled but which have lots of contrary evidence and anecdotal counter stories such as time travel and dimensional crossings.

After you read this book you may question much of what is commonly accepted by scientists in the world today.

Accepted Science & Paradigms Which Are Likely Wrong

Accepted Science & Paradigms Which Are Likely Wrong

Other books by Martin K. Ettington

Spiritual and Metaphysics Books:
Prophecy: A History and How to Guide
God Like Powers and Abilities
Enlightenment for Newbies
Removing Illusions to Find True Happiness
Using the Scientific Method to Study the Paranormal
A Compendium of Metaphysics and How to Guides (Six books together in one volume)
Love from the Heart
The Enlightenment Experience
Learn Your Soul's Purpose
Pursuing Enlightenment
A Modern Man's Search for Truth
Use Intuition and Prophecy to Improve Your Life
The Handbook of Spiritual and Energy Healing
Pure Spirituality and God
Memories Before Birth and Reincarnation
Paranormal Abilities and the Yoga Sutras of Patanjali
Mystical and Magical Societies and Practitioners
Important Prophecies of the Future

Longevity & Immortality:
Physical Immortality: A History and How to Guide
The Commentaries of Living Immortals

Records of Extremely Long Lived Persons
Enlightenment and Immortality
Longevity Improvements from Science
The 10 Principles of Personal Longevity
Telomeres & Longevity
The Diets and Lifestyles of the World's Oldest Peoples
The Longevity Six Books Bundle
Long Lived Plants and Animals
A Guide to Longevity Foods, Diets, and Supplements

Science Fiction:
Out of This Universe

The Immortals of the Interstellar Colony
The Mystic Soldier
The Immortality Sci Fi Bundle

Visiting Many Universes
The History of Science Fiction and Fantasy

The God Like Powers Series:
Human Invisibility
Invulnerability and Shielding
Teleportation
Psychokinesis
Our Energy Body, Auras, and Thoughtforms
The God Like Powers Series—Volume 1 CompilationThe Yoga Discovery Series:
Yoga-An Ancient Art Form
Hatha Yoga-Helping you Live Better
Raja Yoga-Through the Ages
The Yoga Discovery Package

Business & Coaching Books:
Creating, Paublishing, & Marketing Practitioner Ebooks
Building a Successful Longevity Coaching Business
Why Become a Coach?
The Professional Coaching Success Trilogy
2020-Make Money Writing and Selling Books
The 2020 Handbook of High Paying Work Without a College Degree
The important of Creativity and How to Improve Yours
Quantum Mechanics, Technology, Consciousness, and the Multiverse

Self-Improvement
Stress Relief and Methods to do So
The Importance of Creativity and How to Improve Yours

Accepted Science & Paradigms Which Are Likely Wrong

Building Self-Confidence
See the World Clearly
A Trilogy of Self Help Books
A New Paradigm of Truth and Happiness
Building Hope and Wonder Among Chaos
The Importance of Genius In Our World

Science, Technology, and Misc.
Future Predictions By and Engineer & Seer
The Unusual Science & Technology Bundle
Removing Limits On Our Consciousness-And Thinking Outside the Box
Universal Holistic Philosophy
Ball Lightning
Stranger Than Science Stories and Facts
Planet Earth is Conscious

Survival
Survival of Humanity Throughout the Ages
33 Incredible True Survival Stories
The Importance of Fire in History and Mythology
How to Survive Anything: From the Wilderness to Man Made Disasters
Building and Stocking a Nuclear Shelter for less than $10,000
The Human Survival Five Books Bundle
Stranger Than Science Facts and Stories
Stranger Than Science Facts and Stories Volume Two
The Microscopic World Inside and Around Us

Legendary Beings
Are Cryptozoological Animals Real or Imaginary?
Fire in History and Mythology
All About Dragons
Sea Serpents and Ocean Monsters
The Legendary Animals Five Books Bundle
The Mythical People of Ireland
Bigfoot Mysteries and Some Answers
About the Little People: Fairies, Elves, Dwarfs and Leprechauns

Ancient History
The Real Atlantis-In the Eye of the Sahara
Ancient & Prehistoric Civilizations
Ancient & Prehistoric Civilizations- Book Two
The History of Antediluvian Giants
The Antediluvian History of Earth
Ancient Underground Cities and Tunnels
Strange Objects Which Should Not Exist
More Out of Place Artifacts
Strange and Ancient Places in the USA
A Theory of Ancient Prehistory And Giant Aliens
The Destruction of Civilization About 10,500 B.C.
A Timeline of Intelligent Life on Earth
A 300 Million Year Old Civilization Existed on Earth
The Encyclopedia of Out of Place Artifacts
Hollow and Inner Earth Stories and Facts

Aliens and Space
Types of UFOs Observed in History
Aliens and Secret Technology
Aliens Are Already Among Us
Designing and Building Space Colonies
Humanity and the Universe

Living in Space
All About Moon Bases
All About Mars Journeys and Settlement
The Space and Aliens Six Books Bundle
The Space Colonies and Space Structures Coloring Book
All About Asteroids
Spaceships, Past, Present, and Future
Astronauts, Cosmonauts, and Other Important Space Flyers

Accepted Science & Paradigms Which Are Likely Wrong

All About Mars Journeys and Settlement
Mining the Asteroid Belt
The New Era of Space Stations
Moon Landings, Bases, and Exploration

<u>Time Travel and Dimensions</u>
Real Time Travel Stories From a Psychic Engineer
The Real Nature of Time: An Analysis of Physics, Prophecy, and Time Travel Experiences
Stories of Parallel Dimensions
We Live in a Malleable Reality-and We Can Change It
The Time, Dimensions, and Quantum Mechanical Bundle
Alternate Dimensions & the Otherworld
The Multiverse: Time and Dimensional Travel Q&As

<u>Political and Social</u>

The Empire of the United States: Forged By God's Spirit Through Man

The Suppression of Truth in the United States and the World

Accepted Science & Paradigms Which Are Likely Wrong

<u>The Longevity Training Series</u>

(A transcription of the online Multimedia Longevity Coaching Training Program)

The Personal Longevity Training Series-Book1-Long Lived Persons
The Personal Longevity Training Series-Book2-Your Soul's Purpose
The Personal Longevity Training Series-Book3-Enable Your Life Urge
The Personal Longevity Training Series-Book4-Your Spiritual Connection
The Personal Longevity Training Series-Book5-Having Love in Your Heart
The Personal Longevity Training Series-Book6-Energy Body Health
The Personal Longevity Training Series-Book7-The Science of Longevity
The Personal Longevity Training Series-Book8-Physical Body Health
The Personal Longevity Training Series-Book9-Avoiding Accidents
The Personal Longevity Training Series-Book10-Implementing These Principles

The Personal Longevity Training Series-Books One Thru Ten

These books are all available in digital and printed formats from my
website and on Amazon, Barnes & Noble, Apple ITunes, and many other sites

My Books Website is: http://mkettingtonbooks.com

Accepted Science & Paradigms Which Are Likely Wrong

<u>Signup for our Mailing List to get the following:</u>

1) A discount coupon for 25% discount on all books on our site
2) Occasional Notices of new books available
3) Occasional Email on other offerings of ours (Monthly)

If you have any questions about this book or other subjects please contact the Author at:

mke@mkettingtonbooks.com

Accepted Science & Paradigms Which Are Likely Wrong

Accepted Science & Paradigms Which Are Likely Wrong

Table of Contents

1.0 Introduction 1
2.0 Why Accepted Science & Paradigms May Be Wrong 3
3.0 The Big Bang Never Happened 5
   3.1 Specific Evidence Refuting the Big Bang 11
4.0 Can You Travel to Other Dimensions? 19
5.0 Manmade Global Warming 25
   5.1 NOAA's Temperature Data Fraud 25
6.0 Dark Matter Doesn't Exist 31
7.0 Aliens Are Not Proven to Exist 37
   7.1 Evidence Number One-Footprints in Rock .. 37
   7.2 Evidence Number Two-Old Paintings 47
   7.3 Evidence Three-The Dogon Tribe 49
8.0 Ants Can't be Intelligent 53
9.0 Problems with the Standard Model 61
10.0 Life Can't Exist Deep in the Earth 69
11.0 Can People Time Travel? 79
12.0 Civilization On Earth is Recent 95
   12.1 Large Constructions in the Deep Past 95
   12.2 Giants in the Earth 105
   12.3 Civilization 300 Million Years Ago 109
13.0 That Mars Travel Takes Months or Years 113
14.0 How Long Can We Live? 115
15.0 Cryptids (Like Bigfoot) Really Do Exist 121

16.0	Genius is Only for a Few	125
17.0	Little People Are Just a Myth	129
	17.1 Flores Man in Indonesia	129
	17.2 Other Little People Legends	135
18.0	Memories Before Birth	143
19.0	Scientific Method and the Paranormal	149
20.0	Summary	153
21.0	Bibliography	155

Accepted Science & Paradigms Which Are Likely Wrong

1.0 Introduction

As an engineer and student of science I try to keep up with publications of the latest discoveries and paradigm shifts in science around the world.

I believe that there are several big problems with science today. First is that much science is biased which might be because some scientists want to keep their point of view prominent and the others because many focuses of science are supported by biased funding. This will be discussed in this book.

The other issue is that many scientists limit their thinking to only support age old paradigms and refuse to accept changes to their theories or throw out evidence because it doesn't conform their current beliefs.

In this book we will cover many issues such as the belief among many cosmologists that the Big Bang never happened.

We will also look at ideas we thought settled but which have lots of contrary evidence and anecdotal counter stories such as time travel and dimensional crossings.

After you read this book you may question much of what is commonly accepted by scientists in the world today.

Accepted Science & Paradigms Which Are Likely Wrong

Accepted Science & Paradigms Which Are Likely Wrong

2.0 Why Accepted Science & Paradigms May Be Wrong

We constantly need to challenge what we are taught about our world and the Universe. New Scientific ideas are propounded all of the time and some are revolutionary and end up changing the world.

Also, some beliefs are politically based—not based on real science.

In this book we will cover many scientific beliefs which may be wrong or a result of limited thinking.

A good example of a major paradigm shift is plate tectonics. I.E. the idea that the continents move around the world over time and change the layout of our world over hundreds of millions of years.

For many decades in the nineteenth and twentieth centuries geologists believed that changes in the Earth were not sufficient to move continents. The growth of experimental evidence from deep drilling eventually led to a revolution in the 1960s that Tectonic plate movements must occur. Today this is a well-accepted theory of geology.

Scientists become invested in certain views of the world and don't want to change those views even when contrary evidence accumulates. Here are the main reasons why:

A) If a scientist spends many years, does lots of research, and writes lots of papers based on a certain foundation then they are very reluctant to support a point of view which undermines what they believe.

B) Scientific funding also comes from organizations-many of whom have an axe to grind. This is very true in Global Warming "science" since government funded research is very invested in proving Man Made Global Warming to be true. In many cases you can't get a grant to study this question unless you accept the preconceptions.

C) Some potential evidence is so far outside of the accepted norms that it is just rejected or alternative explanations are found. Out of Place Artifacts are a good example. Many of these intelligently made artifacts were found in rock strata which is hundreds of millions of years old. To accept that these are real would be to accept that many theories of anthropology and archeology have huge holes or mistakes in them. So the evidence is just rejected of hand as being "fake".

Accepted Science & Paradigms Which Are Likely Wrong

3.0 The Big Bang Never Happened

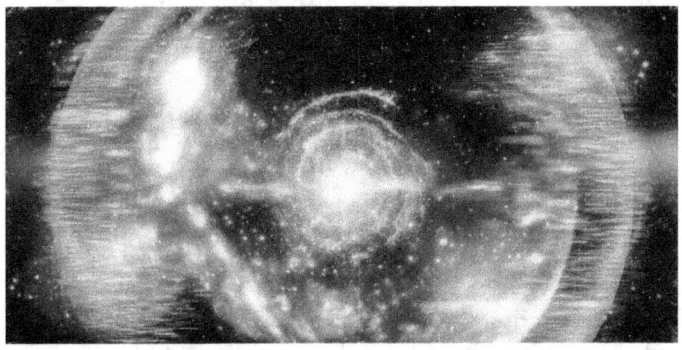

When I grew I was taught that the Big Bang Theory of the creation of the Cosmos was correct and indisputable. Now, scientists are finding out that evidence leads to conclusions that this is not true and we live in a steady state Universe. If this is true it would be an incredible paradigm change.

A tenet of the Big Bang theory, particularly that it produced conditions for certain elements to develop, is on the verge of being dramatically overturned by a scientist who claims evidence shows the event never happened, according to a study.

Under the Big Bang scenario, an explosion occurred at the dawn of our universe 13.8 billion years ago that dispersed chemical elements across space which cooled and formed the galaxies and stars in our cosmos. Modern astronomy's study of the origin and ongoing development of our universe is built largely on the dominant theory's central hypothesis.

But three critical fusion events believed to have been created by the Big Bang are under intense scrutiny by scientist Eric J. Lerner of the nuclear fusion research company LPPFusion.

Scientists believe that precise amounts of helium, deuterium and lithium were produced by fusion reactions in the dense, extremely hot cloud of chemical elements that emerged after the Big Bang.

Lerner, who has spent decades making detailed observations of such reactions, says his and other scientists' findings don't match up with longstanding theories based on observations of older stars. He found that old stars had less than half the helium and less than one tenth the lithium than is predicted by the Big Bang nucleosynthesis theory, which posits that a quarter of the universe's mass is comprised of helium.

According to Lerner – who wrote the book "The Big Bang Never Happened" – no helium or lithium was created before the development of the first stars in our galaxy.

In a statement, Lerner said the mismatch of evidence on the presence of lithium in the cosmos has been well-known for some time among astronomers. But he says challenges to the dominant Big Bang theory – such as the closed-universe and Hubble-constant problems and the failure to find evidence of dark matter – have been dismissed by scientists.

Accepted Science & Paradigms Which Are Likely Wrong

"The Big Bang should have resulted in the annihilation of matter and antimatter, leaving a surviving density of matter that would be a hundred billion times less than that observed," Lerner said in the statement. "To avoid that outcome, Big Bang theory requires an asymmetry of matter and antimatter with consequences, such as the decay of the proton, which have been contradicted by extensive experiments."

In another example, Lerner claims that in a galaxy that is expanding, as the Big Bang theory posits, the surface brightness of distant galaxies should decline over time.

"For cosmology to advance, the basic hypothesis of the Big Bang has to be abandoned," Lerner said in the statement. "The real crisis in cosmology is that the Big Bang never happened."

Lerner says the Galactic Origin of Light Elements, or GOLE, hypothesis, rightly holds that the first generation of stars to form in the cosmos were stars of intermediate mass roughly four to 12 times the size of our sun.

Under the GOLE theory, helium, deuterium and lithium were produced by these stars after they burned hydrogen at faster rates than our sun and dispersed elements across the cosmos through stellar winds.

New observations based on the GOLE hypothesis show that the early stars also produce carbon, boron and beryllium in the amounts observed in the oldest stars.

Accepted Science & Paradigms Which Are Likely Wrong

Lerner said his findings are buttressed by his recent observations of newly formed, more luminous galaxies.

"The correct predictions of the GOLE model not only fit the observations far better than does the Big Bang model" Lerner said in the statement. "The production of the light elements by stars must occur – and if there was also production by a Big Bang, we would observe far more of these light elements than we do."

Not everyone who studies space and the cosmos is ready to get on board with Lerner's theory, however. A Los Angeles-based astronomy and physics professor said longstanding scientific evidence refutes Lerner's claims.

"Many of his arguments don't hold water," University of Southern California professor Vahé Peroomian said in an interview, noting Lerner seldom links to peer-reviewed articles. "My general impression would be to take things he argues with a grain of salt."

Peroomian said that cosmic microwave background, for example, which is evidence of radiation stemming from the Big Bang, is a pillar of the cosmological theory and one that Lerner cannot dispute.

Also, if there were major flaws with the Big Bang theory, Lerner wouldn't be the only critical voice rising from the scientific community, Peroomian said.

Accepted Science & Paradigms Which Are Likely Wrong

Peroomian – who is not a cosmologist – pointed to astrophysicist Edward L. Wright's extensive critique of Lerner's 1991 book, which Peroomian said is part of a chorus of scientific voices taking down Lerner's theories.

Wright, who taught at the University of California, Los Angeles, published an article refuting Lerner's claims that dark matter doesn't exist or that stars contain less helium than the Big Bang predicted.

"Lerner wants to make helium in stars," Wright said in the article. "This presents a problem because the stars that actually release helium back into the interstellar medium make a lot of heavier elements too."

On dark matter, Wright said the evidence for its existence lies in the orbital motions, bending of light and behavior of gases trapped in clusters of galaxies.

Although scientists cannot see dark matter, they can detect it by measuring how its gravity affects stars and galaxies embedded within it.
Using NASA's Hubble Space Telescope, astronomers learned dark matter forms in much smaller clumps around large and medium-size galaxies than previously known.

Accepted Science & Paradigms Which Are Likely Wrong

Accepted Science & Paradigms Which Are Likely Wrong

3.1 Specific Evidence Refuting the Big Bang

The refutation of the Big Bang Theory will have such a major impact on Science, Philosophy, and Religion that we need more evidence of this truth. Here are some specific reasons to support this argument:

1) Light elements: Lithium and Helium

Prediction: Any superhot explosion throughout the universe, like the Big Bang, would have generated a certain small amount of the light element lithium and a large amount of helium.

Observation: Yet as astronomers have observed older and older stars, the amount of lithium observed has gotten less and less, and, in the oldest stars is less than one tenth of the predicted level. The oldest stars near to us have less than half the amount of helium predicted. However, well-understood fusion processes in stars and reactions initiated by cosmic rays have accurately predicted the correct amounts of these and other light elements.

2) Antimatter-matter annihilation

Prediction: Since the intense radiation of the Big Bang would produce matter and antimatter in equal amounts, mutual annulation of particle-antiparticle pairs would reduce the surviving matter density to around 10^{-17} protons/cm3.

Accepted Science & Paradigms Which Are Likely Wrong

Observation: the matter density in the universe is observed to be at least 10 -7 ions /cm3 more than 10 billion times higher than the Big Bang prediction.

Big Bang fix to prediction: To try to fix this well-known vast gap, Big Bang theorists have proposed some unknown asymmetry between matter and antimatter which would lead to more production of matter. This has never been observed in laboratory experiments. A consequence of this predicted imbalance is the decays of the proton, initially predicted to decay with a lifetime of 10 to the 30th years. Large scale experiments have contradicted this prediction was well, with no evidence of decay at all.

3) Surface-Brightness

Prediction: In any expanding universe, an optical illusion makes objects at high redshift appear larger and dimmer, so their surface brightness—the ratio of apparent brightness to apparent area—declines sharply with redshift.

Observation: Based on observations of thousands of galaxies, surface brightness is completely constant with distance, as expected in a universe that is NOT expanding.

Big Bang fix to Prediction: After observations showed that the surface brightness dimming did not occur, Big Bang theorists hypothesized that galaxies were much smaller in the distant past and have grown greatly. But observations have contradicted this fix as well, showing that there have not been enough galaxy mergers for the growth rates needed. In addition, the ultra-small galaxies hypothesized would have to have more mass in stars than total mass, an obvious impossibility.

4) Too Large Structures

Prediction: In the Big Bang theory, the universe is supposed to start off completely smooth and homogenous. Structure starts small and grows over time

Observation: As telescopes have peered farther into space, huger and huger structures of galaxies have been discovered, which are too large to have been formed in the time since the Big Bang.

5) Cosmic Microwave Background Radiation (CMB) and its Anisotropies

Prediction (Initial): The CMB is a smooth relic of the initial radiation of the Big Bang.

Observation: The CMB is smooth on such large scales that in a Big Bang there would be too little time for regions that we now see in different parts of the sky to reach equilibrium with each other, or even to receive energy from each other at the speed of light.

Big Bang fix to prediction: An unknown force, dubbed "inflation" generated an exponential phase of the Big Bang that blew up the universe so rapidly that all asymmetries were smoothed away.

Additional observations: The actual very small anisotropies in the CMB were much smaller than those predicted by Big Bang theorists and additional fixes had to be added to the theory each time the observations became more precise, so that at present seven free variables—the density of dark matter, of ordinary matter, of dark energy and four additional fitting parameters—are needed to fit the observations. They still badly fail with some of the largest-scale anisotropies.

The latest crisis: Based on the data from the Planck satellite, the best fit to the CMB predicts a Hubble constant (the ratio of redshift to distance) in conflict with observations based on Supernovae. The best fits imply a curved universe, in conflict with the predictions of inflation for a flat universe. And they predict a density of dark matter far greater than any measurements derived from the motion of galaxies.

In contrast to the multiple contradictions of the Big Bang theory of the CMB with its "ultra-precise" but wrong predictions, non-Big Bang processes provide a better explanation. The energy that was released in producing the observed helium in the universe equal the energy in the CMB. Any radiation become isotropized if it travels in a medium that scatters it. There is abundant observational evidence that microwave-frequency radiation is scattered in the intergalactic medium.

6) Dark Matter

Prediction: The Big Bang theory requires the existence of dark matter—mysterious particles that have never been observed in the laboratory, despite huge experiments to find them.

Observation: Multiple lines of evidence, especially observations of the motions of galaxies, show that this dark matter does not exist. Extremely sensitive experiments on earth have failed to detect dark matter particles. In addition, dark matter, if it existed would create a viscosity effect on galaxies that would prevent the existence of the many long-lived groups of galaxies that are observed.

The response of most cosmologists to this growing body of evidence has, unfortunately, not been to decide the Big Bang theory has been falsified, but to add new "parameters" and hypotheses, like dark energy. The theory is now far more complex and speculative than the Ptolemaic epicycles that were destroyed by the Scientific Revolution. Each contradiction with observation is taken as a mere "anomaly" that does not undermine the theory as a whole. Strong peer pressure is applied against many of those who question the theory.

"It's as if researchers are saying 'I can see the Emperor's elbow through his New Clothes,' 'I can see the Emperor's knee though his New Clothes' and so on," says Lerner. "It is time to say: 'The Emperor is not wearing any clothes.' This theory has no correct predictions."

To replace the Big Bang, other researchers have elaborated, in peer-reviewed publications, alternative explanations of the generation of light elements and of the energy in the CBR by ordinary stars, and of the development of large-scale structures through the interaction of gravity and electromagnetic processes. "No one would claim that all the problems in cosmology have been resolved," agrees Lerner, "but the evidence is consistent with an evolving, but non-expanding universe, which had no beginning in time and no Big Bang."

7) Galaxies exist which are too old to have been created in the Big Bang.

To everyone who sees them, the new James Webb Space Telescope (JWST) images of the cosmos are beautifully awe-inspiring. But to most professional astronomers and cosmologists, they are also extremely surprising—not at all what was predicted by theory. In the flood of technical astronomical papers published online since July 12, the authors report again and again that the images show surprisingly many galaxies, galaxies that are surprisingly smooth, surprisingly small and surprisingly old. Lots of surprises, and not necessarily pleasant ones. One paper's title begins with the candid exclamation: "Panic!"

Why do the JWST's images inspire panic among cosmologists? And what theory's predictions are they contradicting? The papers don't actually say. The truth that these papers don't report is that the hypothesis that the JWST's images are blatantly and repeatedly contradicting is the Big Bang Hypothesis that the universe began 14 billion years ago in an incredibly hot, dense state and has been expanding ever since. Since that hypothesis has been defended for decades as unquestionable truth by the vast majority of cosmological theorists, the

Accepted Science & Paradigms Which Are Likely Wrong

new data is causing these theorists to panic. "Right now I find myself lying awake at three in the morning," says Alison Kirkpatrick, an astronomer at the University of Kansas in Lawrence, "and wondering if everything I've done is wrong."

Accepted Science & Paradigms Which Are Likely Wrong

4.0 Can You Travel to Other Dimensions?

Physicists have lots of theories about the existence of parallel dimensions but they say there is no evidence of their existence.

There are also legends that the Tuatha Da Dannon who were the ancient race of Ireland came from the "Otherworld" an alternative dimension.

These stories should be seriously examined by scientists since they may be a form of evidence for the existence of other dimensions.

In my book "Stories of Parallel Dimensions" there are stories of persons who seemed to have crossed between dimensions. Here are a couple of example stories:

Four Girls who took a Wrong Turn

Four girls took a wrong turn and found themselves driving in an entirely different, unknown environment.

Accepted Science & Paradigms Which Are Likely Wrong

The ladies had been driving on black asphalt in the desert, but after taking a wrong turn, they said, they found themselves driving on white cement, surrounded by grain field and a lake. They spotted a building with a large, neon sign making up illegible, random squiggles, and as they pulled in for assistance, a large group of tall men poured out of the front door seeming shocked and upset, waving their arms at the girls. Then, the girls realized these "tall men" didn't even appear to be human, so they freaked out and drove off.

While the girls were driving away, they noticed four peculiar, egg-shaped automobiles mounted on tricycle style-wheels were following them. They sped ahead until the mysterious vehicles were out of sight, and when they reached the canyon and drove all the way back through it, they'd somehow returned to the desert they were originally in, glad to be back, but unable to figure out the mysterious place they'd just gone, or how they arrived there.

The Green Children of Woolpit

Accepted Science & Paradigms Which Are Likely Wrong

The title of this story may sound immediately implausible to the cynics amongst you, but surprisingly this is one tale of folklore which is probably founded on some basis of truth!

The legend of the green children of Woolpit starts during the reign of King Stephen, in a rather tumultuous time in England's history called 'The Anarchy' in the mid 12th century.

Woolpit (or in Old English, wulf-pytt) is an ancient village in Suffolk named after – as one might gather from it's name – an old pit for catching wolves! Next to this wolf pit in around 1150, a group of villagers came across two young children with green skin, apparently speaking gibberish and acting nervously.

According to contemporary writings at the time by Ralph of Coggeshall, the children were subsequently taken to the nearby home of Sir Richard de Calne where he offered them food but they repeatedly refused to eat. This continued for some days until the children came across some green beans in Richard de Calne's garden which they ate straight out of the ground.

It is thought that the children lived with Richard de Calne for some years, where he was able to slowly convert them over to normal food. According to the writings of the day, this change in diet led to the children losing their green complexion.

The children also slowly learnt to speak English, and once fluent were asked where they had come

Accepted Science & Paradigms Which Are Likely Wrong

from and why their skin was once green. They replied with:

"We are inhabitants of the land of St. Martin, who is regarded with peculiar veneration in the country which gave us birth."

"We are ignorant [of how we arrived here]; we only remember this, that on a certain day, when we were feeding our father's flocks in the fields, we heard a great sound, such as we are now accustomed to hear at St. Edmund's, when the bells are chiming; and whilst listening to the sound in admiration, we became on a sudden, as it were, entranced, and found ourselves among you in the fields where you were reaping."

"The sun does not rise upon our countrymen; our land is little cheered by its beams; we are contented with that twilight, which, among you, precedes the sun-rise, or follows the sunset. Moreover, a certain luminous country is seen, not far distant from ours, and divided from it by a very considerable river."

Shortly after this revelation Richard de Calne took the children to be baptized in a local church, however the boy died soon afterwards through an unknown illness.

The girl, later known as Agnes, continued to work for Richard de Calne for many years before marrying the archdeacon of Ely, Richard Barre. According to one report, the pair had at least one child.

Accepted Science & Paradigms Which Are Likely Wrong

So who were the green children of Woolpit?

The most likely explanation for the green children of Woolpit is that they were the descendants of Flemish immigrants who had been persecuted and possibly killed by King Stephen or – perhaps – King Henry II. Lost, confused and without their parents, the children could have ended up at Woolpit speaking only their native tongue of Flemish, perhaps explaining how the villagers thought that they were speaking gibberish.

Furthermore, the green tint to the children's skin could be explained by malnourishment, or more specifically 'green sickness'. This theory is supported by the fact that their skin reverted to a normal colour once Richard de Calne had converted them over to eating real food.

Personally, we like to side with the more romantic theory that these children arrived from an underground world where the native inhabitants are all green!

Accepted Science & Paradigms Which Are Likely Wrong

Accepted Science & Paradigms Which Are Likely Wrong

5.0 Manmade Global Warming

There are lots of reasons to say that Manmade Global Warming is a fraud.

One of these issues is that satellite measurements of the world's temperature haven't changed in eighteen years. Here is a quote on the matter:

The Earth's temperature has "plateaued" and there has been no global warming for at least the last 18 years, says Dr. John Christy, professor of atmospheric science and director of the Earth System Science Center (ESSC) at the University of Alabama/Huntsville.

What about NOAAs fraudulent temperature records:

5.1 NOAA's Temperature Data Fraud

The National Oceanic and Atmospheric Administration (NOAA) may have a boring name, but it has a very important job: It measures U.S. temperatures. Unfortunately, it seems to be a captive of the global warming religion. Its data are fraudulent.

What do we mean by fraudulent? How about this:

NOAA has made repeated "adjustments" to its data, for the presumed scientific reason of making the data sets more accurate.

Nothing wrong with that. Except, all their changes point to one thing — lowering previously measured temperatures to show cooler weather in the past, and raising more recent temperatures to show warming in the recent present.

This creates a data illusion of ever-rising temperatures to match the increase in CO_2 in the Earth's atmosphere since the mid-1800s, which global warming advocates say is a cause-and-effect relationship. The more CO_2, the more warming.

But the actual measured temperature record shows something different: There have been hot years and hot decades since the turn of the last century, and colder years and colder decades. But the overall measured temperature shows no clear trend over the last century, at least not one that suggests runaway warming.

That is, until the NOAA's statisticians "adjust" the data. Using complex statistical models, they change the data to reflect not reality, but their underlying theories of global warming. That's clear from a simple fact of statistics: Data generate random errors, which cancel out over time. So by averaging data, the errors mostly disappear.

That's not what NOAA does.

According to the NOAA, the errors aren't random. They're systematic. As we noted, all of their temperature adjustments lean cooler in the distant past, and warmer in the more recent past. But they're very fuzzy about why this should be.

Accepted Science & Paradigms Which Are Likely Wrong

Far from legitimately "adjusting" anything, it appears they are cooking the data to show a politically correct trend toward global warming. Not by coincidence, that has been part and parcel of the government's underlying policies for the better part of two decades.

What NOAA does aren't niggling little changes, either.

As Tony Heller at the Real Climate Science web site notes, "Pre-2000 temperatures are progressively cooled, and post-2000 temperatures are warmed. This year has been a particularly spectacular episode of data tampering by NOAA, as they introduce nearly 2.5 degrees of fake warming since 1895."

So the global warming scare is basically a hoax. This winter, for instance, as measured by temperature in city after city and by snow-storm severity, has been one of the coldest on record in the Northeast.

But after the NOAA's wizards finished with the data, it was merely about average.

Climate analyst Paul Homewood notes for instance that in New York state, measured temperatures this year were 2.7 degrees or more colder than in 1943. Not to NOAA. Its data show temperatures this year as 0.9 degrees cooler than the actual data in 1943.

Accepted Science & Paradigms Which Are Likely Wrong

Erasing Winter

By the way, a similar result occurred after the brutally cold 2013-2014 winter in New York. It was simply adjusted away. Do this year after year, and with the goal of radically altering the temperature record to fit the global warming narrative, and you have what amounts to climate fraud.

"Clearly NOAA's highly homogenized and adjusted version of the Central Lakes temperature record bears no resemblance at all the actual station data," writes Homewood. "And if this one division is so badly in error, what confidence can there be that the rest of the U.S. is any better?"

That's the big question. And for those who think that government officials don't have political, cultural or other agendas, that's naiveté of the highest sort. They do.

Since the official government mantra for all of the bureaucracies at least since the Clinton era is that CO_2 production is an evil that inevitably leads to runaway global warming, those who toil in the bureaucracies' statistical sweat shops know that their careers and future funding depend on having the politically correct answers — not the scientifically correct ones.

"The key point here is that while NOAA frequently makes these adjustments to the raw data, it has never offered a convincing explanation as to why they are necessary," wrote James Delingpole recently in Breitbart's Big Government. "Nor yet,

Accepted Science & Paradigms Which Are Likely Wrong

how exactly their adjusted data provides a more accurate version of the truth than the original data."

There are at least some signs of progress, however. In the case of the Environmental Protection Agency, future reports and studies will include the data and the underlying scientific assumptions for public scrutiny.

That's one way to bring greater honesty to government — and to keep climate charlatans from bankrupting our nation with spurious demands for carbon taxes and deindustrialization of our economy to prevent global warming. The only real result won't be a cooler planet, but rather mass poverty and lower standards of living for all.

Accepted Science & Paradigms Which Are Likely Wrong

6.0 Dark Matter Doesn't Exist

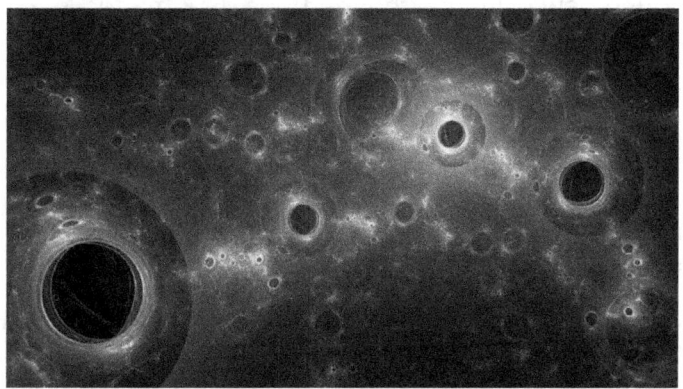

Maybe 'dark matter' doesn't exist after all, new research suggests.

Observations of distant galaxies have seen signs of a modified theory of gravity that could dispense with the invisible, intangible and all-pervasive dark matter.

For decades, astronomers, physicists and cosmologists have theorized that the universe is filled with an exotic material called "dark matter" that explains the stranger gravitational behavior of galaxies and galaxy clusters.

Dark matter, according to mathematical models, makes up three-quarters of all the matter in the universe. But it's never been seen or fully explained. And while dark matter has become the prevailing theory to explain one of the bigger mysteries of the universe, some scientists have looked for alternative explanations for why galaxies act the way they do.

Accepted Science & Paradigms Which Are Likely Wrong

Now, an international team of scientists says it has found new evidence that perhaps dark matter doesn't really exist after all.

In research published in November in the Astrophysical Journal, the scientists report tiny discrepancies in the orbital speeds of distant stars that they think reveals a faint gravitational effect – and one that could put an end to the prevailing ideas of dark matter.

The study suggests an incomplete scientific understanding of gravity is behind what appears to be the gravitational strength of galaxies and galaxy clusters, rather than vast clouds of dark matter.

That might mean pure mathematics, and not invisible matter, could explain why galaxies behave as they do, said study co-author Stacy McGaugh, who heads the astronomy department at Case Western Reserve University in Cleveland.

The new research reports that signs of a faint gravitational tide, known as the "external field effect" or EFE, can be observed statistically in the orbital speeds of stars in more than 150 galaxies.

The authors say the effect cannot be explained by dark matter theories, but it's predicted by what's known as the modified Newtonian dynamics theory, or MOND.

"What we're really saying is that there is absolutely evidence for a discrepancy," McGaugh said. "What you see is not what you get, if all you know about is Newton and Einstein."

Accepted Science & Paradigms Which Are Likely Wrong

Astronomers long assumed that stars orbited the centers of galaxies at speeds predicted by the theory of gravity formulated by the English physicist and mathematician Isaac Newton more than 300 years ago.

Newton based his theory that objects attract each other with a force varying according to their mass on observations of the orbits of the planets. With refinements from the theories of the German-born physicist Albert Einstein in the 20th century, it remains astonishingly accurate.

But observations of the Coma cluster of galaxies in the 1930s by Swiss astronomer Fritz Zwicky, then working at the California Institute of Technology, found it was subject to larger-than-expected gravitational forces – an effect he attributed to "dunkel (kalt) materie," which is German for "dark (cold) material."

When the American astronomers Vera Rubin and Kent Ford found anomalies in the orbits of stars in galaxies in the 1970s, many scientists theorized they were caused by masses of invisible "dark matter" within and around galaxies, and the idea has dominated astrophysics ever since.

By some estimates, dark matter makes up about 85 percent of all the matter in the universe. It's said to interact with light and visible matter only through gravity, and it explains the observed anomalies in distant galaxies.

Accepted Science & Paradigms Which Are Likely Wrong

MOND was formulated in the 1980s by an Israeli physicist, Mordehai Milgrom, to explain the observed discrepancies without dark matter.

It proposes that gravity causes a very small acceleration, not predicted by Newton and Einstein, at such low levels that it can only be seen in galaxy-size objects; and it would mean the explanation of dark matter is not needed.

So far, MOND has survived several scientific tests – although many scientists say it cannot explain observations of the Bullet cluster of colliding galaxies, for example.

McGaugh admits that MOND is a minority view in astrophysics, and that most scientists favor the existence of dark matter – an idea he favored himself, until he began to change his mind about 25 years ago.

"I once would have said the same things: it's absolutely proven that there's dark matter, don't worry about it," he said.

But many of the predictions of MOND have been seen in astronomical observations, and the latest research is one more piece of evidence for it, he said.

"MOND is the only theory that has succeeded in this way," McGaugh said. "It is the only theory that has routinely had all predictions come true."

The new research raises "a very interesting issue," said Matthias Bartelmann, a professor of theoretical astrophysics at Heidelberg University in Germany, who was not involved in the study.

"Can dark matter be explained by a different law of gravity? It would be most important for cosmology as well as particle physics if it could," he said in an email.

He has doubts, however, that the "external field effect" reported in the new research is truly a unique prediction of MOND, and that it cannot be explained by some competing theories.

And since MOND theory was formulated to account for the rotational discrepancies in galaxies, testing it on galaxies would be expected to return convincing results; instead, MOND needed to be tested successfully on other objects, such as galaxy clusters, he said.

Accepted Science & Paradigms Which Are Likely Wrong

Accepted Science & Paradigms Which Are Likely Wrong

7.0 Aliens Are Not Proven to Exist

I've written six books on UFOS and Aliens and I contradict the belief that Aliens are not proven to exist since my research shows that not only do they exist but they have been on Earth for thousands if not millions of years.

Here is just a little bit of the evidence from my book "Four Evidences for Aliens and UFOs in Earth's History". (Only Three Types of evidence covered here)

7.1 Evidence Number One-Footprints in Rock

Giants existed far in Earth's past who could not be related to Modern Man. They were likely Aliens or created by Aliens.

Human & Dino Footprints-100 Million Years Old

Alvis Delk Print
in the Sir George Series

Accepted Science & Paradigms Which Are Likely Wrong

Over the years a large number of fossilized human tracks have been reported at various locations around the world. Some of these shed light on the coexistence of men and dinosaurs.

The Paluxy River basin in Glen Rose Texas is the location of Dinosaur Valley State Park. Many dinosaur tracks have been found along the river and a large number have been excavated to preserve them from erosion. But there have also been human tracks found in this same rock layer. To the right is the Willet print, which was excavated from a limestone ledge near Dinosaur Valley State Park.

Below to the left is the Feminine Print, a "human track inside a dinosaur track," that was found in the Paluxy River area of Glen Rose, Texas. In the center is the Delk Print, which shows a human footprint intruded by a tridactyl dinosaur print. The Delk Track has been authenticated by spiral CT scan, which can verify that there is greater compression density below the tracks then elsewhere in the rock. The right picture shows what are called "following contours" revealed by the CT scan. These would not be there if the track was carved. These Paluxy "man-tracks alongside dinosaur-tracks" have been the source of considerable controversy over the years.

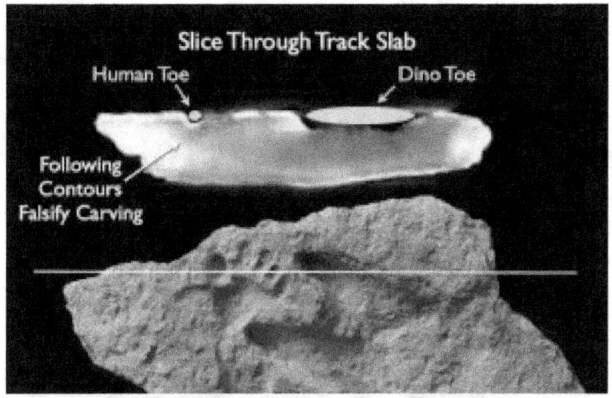

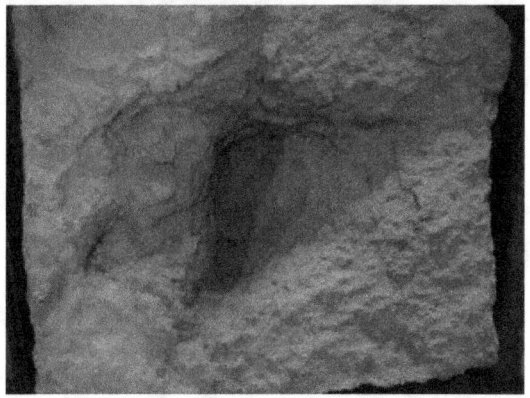

Originally the Paluxy ichnofossils (or trace fossils) were considered by creationists to be powerful evidence that men and dinosaurs coexisted. In the 1980s John Morris wrote the popular book *Tracking Those Incredible Dinosaurs (and the People Who Knew Them)* and the film "Footprints in Stone" was produced by Stan Taylor.

Over time, the exposed prints became quite eroded and evolutionists argued that they were merely elongated dinosaur footprints that had experienced infilling. Some creationists are now inclined to agree that the famous Taylor Trail was made by a dinosaur, though some point to the mixture of human and dinosaur characteristics as evidence that the tracks are a composite, the human track superimposed upon the existing dinosaur footprints. (See Robert Helfinstine and Jerry Roth's 1994 book *Texas Tracks and Artifacts*.)

The lack of clarity regarding these original Paluxy "man-tracks" finds prompted leading creationists to cease using the Paluxy footprints as evidence for men living with dinosaurs. But then additional tracks, like the Feminine Print and the Delk Track, came to light, providing much clearer evidence. It is instructive to consider that these Paluxy human footprints are much more distinct than Mary

Accepted Science & Paradigms Which Are Likely Wrong

Leakey's famous Laetoli Track in Tanzania (left), which is universally accepted as hominid.

The limestone beds of the Paluxy River containing the supposed human and dinosaur footprints are thought by evolutionists to be 120 million years old. Milne and Schafersman admit, "Such an occurrence, if verified, would seriously disrupt conventional interpretations of biological and geological history and would support the doctrines of creationism and catastrophism." (Milne, and Schafersman, "Dinosaur Tracks, Erosion Marks and Midnight Chisel Work (But No Human Footprints) in the Cretaceous Limestone of the Paluxy River Bed, Texas," *Journal of Geological Education*, Vol. 31, 1983, pp. 111-123.)

Incidentally, the Laetoli prints are also problematic for evolutionists because they appear fully modern and yet the rock layer is dated to 3.5-3.7 million years ago, too old for modern Homo sapiens in the current paradigm of human evolution.

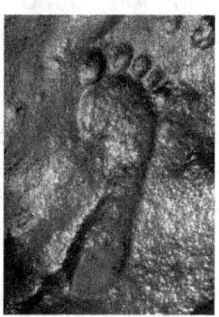

To the right is pictured the Zapata track, found in Permian limestone in New Mexico. The Permian is thought by evolutionary geologists to be over 250 million years old. Yet there is a clear fossil human footprint. It is a very shallow track, almost invisible unless wet with strong side lighting. This accounts for the dramatic hour glass shape with dots in front, similar to what you see when you walk with a wet foot on a tile floor.

Geologist Don Patton attempted to cut this print out of the rock, but wore out four carborundum blades trying to make one cut! Patton reports to have personally seen a photograph of four, virtually identical tracks in an obvious right left pattern taken about one quarter mile from the Zapata track. The rock and the tracks look virtually identical. Some critics claim the Zapata print is "too perfect." But the mud push-up on the sides and the fact that the matrix proved extremely hard to cut out (lab tests indicated it was limestone with 30% silica) would make a carving quite unlikely.

In 1987, not far from the Zapata track site, paleontologist Jerry MacDonald discovered a variety of beautifully preserved fossil footprints in Permian strata. The Robledo Mountain site contains thousands of footprints and invertebrate trails that represent dozens of different kinds of animals. Because of the quality of preservation and sheer multitude of different kinds of footprints, this tracksite has been called one of the most important Early Permian sites ever discovered. Some that have visited the site remark that it contains what appears to be a barefoot human print. "The fossil tracks that MacDonald has collected include a number of what paleontologists like to call 'problematica.'

Accepted Science & Paradigms Which Are Likely Wrong

On one trackway, for example, a three-toed creature apparently took a few steps, then disappeared–as though it took off and flew. 'We don't know of any three-toed animals in the Permian,' MacDonald pointed out. 'And there aren't supposed to be any birds.' He's got several tracks where creatures appear to be walking on their hind legs, others that look almost simian. On one pair of siltstone tablets, I notice some unusually large, deep and scary-looking footprints, each with five arched toe marks, like nails. I comment that they look just like bear tracks. 'Yeah,' MacDonald says reluctantly, 'they sure do.' Mammals evolved long after the Permian period, scientists agree, yet these tracks are clearly Permian." ("Petrified Footprints: A Puzzling Parade of Permian Beasts," *The Smithsonian*, Vol. 23, July 1992, p.70.)

To the left is the "Meister Print," found in Utah within a block of shale. It was first publicized in the *CRS Quarterly* as the footprint containing a trilobite fossil. Bottom left is a fossilized shoe sole found petrified in Triassic rock. This print specimen is so clear that the threads are visible to the naked eye! Also published in this journal is the 1995 study of quasihuman ichnofossils (supposed human tracks) found with tracks of dinosaurs in strata near Tuba City, Arizona.

Photomicrographic analysis indicates that the human-like impressions were created by pressure which created relatively smooth surfaces, unlike the rougher surfaces of impressions formed inside concretions and unlike surrounding surfaces. Comparison of the quasihuman ichnofossils with modern tracks in wet mud shows them to be closely comparable, supporting their theory that the fossil imprints were made by human feet. (Auldaney, Rosnau, Back, and Davis, *CRS Quarterly*, vol. 34, pp. 133-146.)

In 1983 Professor Amanniyazov, Director of Turkmenia's Institute of Geology, reported what appeared to be human footprints in Mesozoic strata. "This spring, an expedition from the Institute of Geology of the Turkmen SSR Academy of Sciences led by found over 1,500 tracks left by dinosaurs in the mountains in the south-east of the Republic. Impressions resembling in shape a human footprint were discovered next to the tracks of the prehistoric animals." (Rubstsov, "Tracking Dinosaurs," *Moscow News*, No. 24, p. 10, 1983.) Dr. Amanniqazov was shocked beyond belief to find a human footprint mingled with dinosaurs. He discusses one of the footprints and says: "if we speak of the human footprint, it was made by a human or a human-like animal. Incredibly, this footprint is on the same plateau where there are dinosaur tracks. We can say the age of this footprint is not 5 or 10, but at least 150 million years old. It is 26cm long, that is Russian size 43 EEE [9.5 American], and we consider that whoever left the footprint was taller than we are...this would create a revolution in the science of man."

(Amanniyazov, Kurban, Science in the USSR T 986, "Old Friends Dinosaurs," p. 103-107.) There is also this fascinating quote from the Russian journalist, Alexander Bushev who investigated these trackways: "But the most mysterious fact is that among the footprints of dinosaurs, footprints of bare human feet were found...We know that humans appeared much later than dinosaurs – that there was an extraterrestrial who walked in his swimming suit along the sea side." (Bushnev, Alexander, Komsomolskya Pravda, January 31, 1995, p. 61ff.)

Perhaps the most intriguing such fossil footprint report was that made by the head of department at Berea College in Kentucky of a human-like track left in sandstone of the Upper Carboniferous Period. Numerous scientists have investigated these tracks and concluded that they are genuine (even going so far as to count the sand grains under magnification to ensure that it was compressed at the bottom rather than carved).

In *Scientific American*, geologist Albert G. Ingalls writes, "If man, or even his ape ancestors, or even that ape ancestor's early mammalian ancestor, existed as far back as the Carboniferous Period in any shape, then the whole science of geology is so completely wrong that all the geologists will resign their jobs and take up truck driving. Hence, for the present at least, science rejects the attractive explanation that man made these mysterious prints in the mud of the Carboniferous with his feet." Ingalls suggested that they were made by some unidentified amphibian. But a human-sized Carboniferous amphibian is just about as problematic for evolutionary timetables as humans in that era!

However, in an attempt to dismiss these tracks, the *Scientific American* article did not include the real photos in their article, instead showing some pretty obvious fakes (probably Indian carvings) and not the actual prints, which they had access to. Why would they not show the real tracks? Because this evidence is highly problematic to their worldview, the theory of evolution. As evolutionary atheist Richard Dawkins observed, authenticated evidence of humans in the Carboniferous would "blow the theory of evolution out of the water." (Dawkins, *Free Inquiry*, vol. 21, no. 4, 2001.)

Accepted Science & Paradigms Which Are Likely Wrong

7.2 Evidence Number Two-Old Paintings

There are many old paintings from hundreds of years ago which show UFOs. Here is one of them:

The Annunciation with Saint Emidius 1486 A.D.

The Annunciation, with Saint Emidius is an altarpiece by Italian artist Carlo Crivelli showing an artistic adaptation of the Annunciation. The altarpiece was painted for the Church of SS. Annunziata in the Italian town of Ascoli Piceno, in the region of Marche, to celebrate the self-government granted to the town in 1482 by Pope Sixtus IV. The painting was removed to the Pinacoteca di Brera in Milan in 1811, but passed to Auguste-Louis de Sivry in 1820, and had reached

Accepted Science & Paradigms Which Are Likely Wrong

England by the mid-19th century. It has been housed in the National Gallery in London since it was donated by Henry Labouchere, 1st Baron Taunton in 1864.

Note the UFO shooting an energy ray to the building in the middle of the painting. Here is a zoom into the painting showing the UFO and energy ray in more detail:

Note that the ray goes also goes into the building and then into the head of the young woman:

7.3 Evidence Three-The Dogon Tribe

In Mali, West Africa, lives a tribe of people called the Dogon. The Dogon are believed to be of Egyptian decent and their astronomical lore goes back thousands of years to 3,200 BC.

According to their traditions, the star Sirius has a companion star which is invisible to the human eye. This companion star has a 50 year elliptical orbit around the visible Sirius and is extremely heavy. It also rotates on its axis.

This legend might be of little interest to anybody but the two French anthropologists, Marcel Griaule and Germain Dieterlen, who recorded it from four Dogon priests in the 1930's. Of little interest except that it is exactly true.

How did a people who lacked any kind of astronomical devices know so much about an invisible star? The star, which scientists call Sirius

Accepted Science & Paradigms Which Are Likely Wrong

B, wasn't even photographed until it was done by a large telescope in 1970.

The Dogon stories explain that also. According to their oral traditions, a race of people from the Sirius system called the Nommos visited Earth thousands of years ago. They also appear in Babylonian, Accadian, and Sumerian myths. The Egyptian Goddess Isis, who is sometimes depicted as a mermaid, is also linked with the star Sirius.

The Nommos, according to the Dogon legend, lived on a planet that orbits another star in the Sirius system. They landed on Earth in an "ark" that made a spinning decent to the ground with great noise and wind. It was the Nommos that gave the Dogon the knowledge about Sirius B.

It doesn't seem to explain a 400-year old Dogon artifact that apparently depicts the Sirius configuration nor the ceremonies held by the Dogon since the 13th century to celebrate the cycle of Sirius A and B. It also doesn't explain how the Dogons knew about the super-density of Sirius B, a fact only discovered a few years before the anthropologists recorded the Dogon stories.

The Dogons are a people well known by their cosmogony, their esotericism, their myths and legends that interest foreigners at the highest point in search for culture or tourism. The population is assessed to be about 300,000 people living in the South West of the Niger loop in the region of Mopti in Mali (Bandiagara, Koro, Banka), near Douentza and part of the North of Burkina (North west of Ouahigouya).

Accepted Science & Paradigms Which Are Likely Wrong

The Dogon's (Mali, Africa) homeland has been designated a World Heritage site for its cultural and natural significance. They are also famous for their artistic abilities and vast knowledge about astrology, especially the Sirius star, which is the center of their religious teachings. The Dogons know that Sirius A, the brightest system in our firmament, is next to a small white dwarf called Sirius B, which was not identified by western scientists until 1978. The Dogons knew about it at least 1000 years ago. Sirius B has formed the basis of the holiest Dogon beliefs since antiquity.

Western astronomers did not discover the star until the middle of the nineteenth century, and it wasn't even photographed until 1970. The Dogons go as far as describing a third star in the Sirius system, called "Emme Ya" that, to date, has not been identified by astronomers.

In addition to their knowledge of Sirius B, the Dogon mythology includes Saturn's rings and Jupiter's four major moons. They have four calendars, for the Sun, Moon, Sirius, and Venus, and have long known that planets orbit the sun.

Accepted Science & Paradigms Which Are Likely Wrong

8.0 Ants Can't be Intelligent

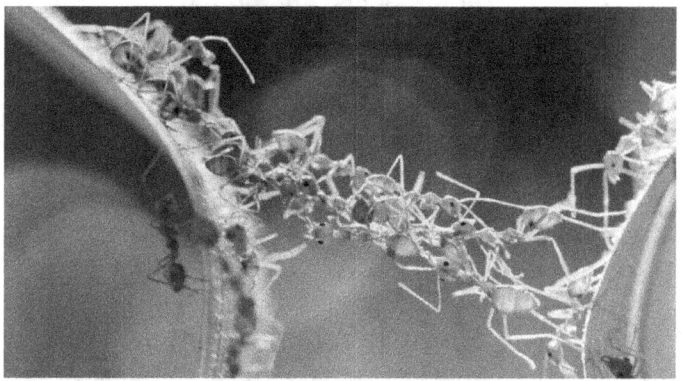

Ants don't have brains like mammals do so they can't be intelligent. Right? Not so fast. There is a lot of evidence that they have some type of group intelligence.

Maybe they have a different type of intelligence than we do. They might have some type of communal group intelligence and communicate by means we don't know-like telepathy.

Just look at these facts:

Intelligent Ants

Is it possible that ants are intelligent? The idea may seem preposterous to some—after all, how can something so small, with the brain the size of a pin head be smart? The very thought of bugs and insects being intelligent seems like an insult to us humans. After all, aren't we the dominant species; the only species that builds cities, uses tools, farms, and demonstrates the capacity to plan and think?

Accepted Science & Paradigms Which Are Likely Wrong

But if we look closer, we can see that ants exhibit many of the characteristics and behaviors that we associate with intelligence and civilization. In fact, if ants did not exist on Earth but we encountered them on, for example, Mars, I am sure that we would wonder if we had encountered an intelligent alien race that builds cities, farms, raises animals, and organizes itself into a complex society complete with social ranks such as nobles, soldiers, workers and slaves. I am sure that we would conclude that these aliens were in fact intelligent. So why do we ignore the signs of intelligence of ants on our own world? Do we have an intelligent alien species literally here under our feet?

So let's explore the alien world of ants right here on Earth and see whether they are intelligent or not.

Ants Build Cities

I know what you are thinking, ant hills aren't cities. They're, well, ant hills. But did you know that large ant hills contain complex ventilation systems that remove carbon dioxide and bring in fresh air, or that they have the equivalent of hundreds of miles of sewers that drain the ant waste into special chambers were the waste is recycled? Did you know that ant cities have an incredibly complex transportation system including highways? Or that each ant city can hold millions of ants.

Sounds incredible, and for the most part it is difficult to imagine the engineering marvel which is an ant city because most of it is underground. In fact, if we were the size of an ant, most of an ant city would be the equivalent of three miles underground.

Accepted Science & Paradigms Which Are Likely Wrong

Ants Farm and Cultivate Mushrooms

Ants are the only animal besides humans that farms food. All other creatures hunt or harvest their food where they find it and are dependent on the whims of nature, and climate for their survival. For example, wolves are smart, and they will exhibit cooperation and skill in hunting for food. But wolves do not capture deer and breed them. Deer will forage for grasses and other food, but of course they have no thought of sowing grass seeds to ensure a plentiful supply of foraging crops. In fact, not one animal besides man and ants has ever thought to keep their prey in captivity or to farm plants in order to feed themselves in the future. Even intelligent animals like wolves lack the foresight to plan beyond meeting their immediate needs.

Ants, like humans, farm plants and raise cattle. Sounds preposterous. It's true.

There are species of ants that collect leaves and take them to specially constructed chambers within their colonies where they grow fungus on the decomposing leaves. The fungus is then eaten by the ants.

The growing of the fungus requires a great deal of planning and forethought: an appropriate chamber must be constructed, the right leaves must be collected, waste must be removed so as not to choke the growing fungus beds, and the leaves must be seeded with the fungus spores. The spores do no grow naturally in throughout the ant colony; the ants must collect the spores and bring them to the leaves.

Fungus farming is an example of intelligence and creativity. Other animals and insects would recognize the food value of fungus growing on leaves if they came across it in the wild. But no other animal or insect, besides humans, would understand that by contaminating a new leaf with the fungus spore, it will result in more food later. This shows intelligence, understanding and the ability to think ahead.

The fact that ants farm is an achievement that sets them apart from the rest of the animal and insect kingdoms. What is even more amazing is that ants have been doing this for millions of years. Humans did not learn to farm until around 5 or 6,000 years ago. Prior to that, humans behaved as hunter gatherers just like the rest of the animal kingdom.

Ants Farm Other Insects

But ants don't just farm, they raise and keep other insects for food, just like humans raise cattle. Many species of ants will domesticate aphids and act like shepherds by taking the aphids to feed on plants, while protecting them from other insect predators. The ants will then "milk" the aphids by squeezing their abdomens and causing some digested plant juice to be released into the mouths of the ants which will then share this nutritious fluid with the rest of the colony.

The ant's behavior in keeping ants closely parallels that of human shepherds and cattle breeders: ants will take the aphids to different pastures, they will guard them against predators, and they will harvest them.

The ants' behavior in this regard is markedly different from that of other animals or insects. Even though wolves display intelligence similar to that of dogs, they lack the foresight to control their instincts and avoid killing their prey in order to get more food in the long run. If a wolf gets his teeth on a rabbit, or a deer, it will kill it and eat it on the spot. No wolf would ever capture the animal, tend to its needs, protect it from other predators and then take food from it without killing it (for example milking a cow) in order to reuse this food resource.

The only animals that do this are humans and ants. And once again ants beat us to it: they have been farming aphids for millions of years. Humans discovered animal husbandry about 6,000 years ago.

Ants Wage War

Ants are the only animal besides humans which wage war in organized battalions, against other organized opponents. Like humans, ants wage war to capture territory and food resources from other ant colonies. Sometimes ant wars lead to the total defeat of an opponent and the survivors are captured and held as slaves.

Of course, war in itself may not be a great example of intelligence. But the organization, planning and coordination required to wage war is the product of intelligence.

In contrast to the war waging behavior of many ant colonies, some ant species settle their difference in single combat between champions chosen by each colony. Bert Holldobler, in an article entitled Tournaments and Slavery in a Desert Ant, noted that a species of desert ant conducts tournaments "in which hundreds of ants perform highly stereotyped display fights". The losing ant colony is then enslaved.

Ants Capture Slaves

Ant wars will often result in the defeated survivors being kept as slaves by the victorious ant colony. They are incorporated into the new colony and made to work for the victors.

We must not equate ant slavery with the human experience. Obviously human slavery is morally reprehensible and wrong from a political, moral and economic perspective. Still, the taking of prisoners and using them as slaves is a behavior that is both complex and unique to ants and humans.

When other animals defeat a foe, they either kill it or allow it to retreat. For example, if two male mountain goats fight over a female, they will ram their horns against each other until one either dies or retreats. If the loser retreats, the winner will win right to mate with the female goat. No animal would then make the loser his slave.

Ants, on the other hand, have figured out that defeated enemies can be useful. They can be spared and put to work for the good of the colony.

Accepted Science & Paradigms Which Are Likely Wrong

The ants' behavior in capturing and enslaving other ants shows an understanding of 1) deferred benefit (it is better to use the slave ants for future work than to eat them now) and 2) organization (slave ants must be supervised and put to work on assigned tasks).

Ants Teach and Communicate

A recent study has demonstrated that ants can pass on knowledge from one ant to another and teach other ants how to find food.

Ants have been observed to use a teaching technique called "tandem running" in which an ant that knows where to find food, will lead a new ant to the spot. The teacher ant will slow his pace to allow the student ant; if the student ant falls behind.

The teacher ant's behavior does not provide a benefit to the teacher. If the teacher were not leading the student ant, it could locate and collect the food about four times faster. But by taking time to lead a novice ant to a food source, it allows other ants to locate the food faster than they would have discovered it on their own. As a result, the entire ant nest benefits.

Scientists believe that this ant behavior represents "the first time a demonstration of formal teaching has been recognized in any non-human animal". Once again, humans and ants have something in common.

Accepted Science & Paradigms Which Are Likely Wrong

Ants Cooperate and Exhibit Teamwork

Ants are tiny, but they can cooperate to an amazing degree. Their cooperation exhibits purpose, planning, and command and control.

Their behavior parallels that of humans. Imagine an ancient workforce of Egyptian laborers building the pyramids by moving giant limestone blocks, and you will have a good comparison to the amazing ants.

There is a lot more we could cover on Ant intelligence but there is lots of supporting research out there.

9.0 Problems with the Standard Model

Five mysteries the Standard Model can't explain. Our best model of particle physics explains only about 5 percent of the universe.

The Standard Model is a thing of beauty. It is the most rigorous theory of particle physics, incredibly precise and accurate in its predictions. It mathematically lays out the 17 building blocks of nature: six quarks, six leptons, four force-carrier particles, and the Higgs boson. These are ruled by the electromagnetic, weak and strong forces.
"As for the question 'What are we?' the Standard Model has the answer," says Saúl Ramos, a researcher at the National Autonomous University of Mexico (UNAM). "It tells us that every object in the universe is not independent, and that every particle is there for a reason."

For the past 50 years such a system has allowed scientists to incorporate particle physics into a single equation that explains most of what we can see in the world around us.

Despite its great predictive power, however, the Standard Model fails to answer five crucial questions, which is why particle physicists know their work is far from done.

Accepted Science & Paradigms Which Are Likely Wrong

1. Why do neutrinos have mass?

Three of the Standard Model's particles are different types of neutrinos. The Standard Model predicts that, like photons, neutrinos should have no mass.

However, scientists have found that the three neutrinos oscillate, or transform into one another, as they move. This feat is only possible because neutrinos are not massless after all.

"If we use the theories that we have today, we get the wrong answer," says André de Gouvêa, a professor at Northwestern University.

The Standard Model got neutrinos wrong, but it remains to be seen just how wrong. After all, the masses neutrinos have are quite small.

Is that all the Standard Model missed, or is there more that we don't know about neutrinos? Some experimental results have suggested, for example, that there might be a fourth type of neutrino called a sterile neutrino that we have yet to discover.

2. What is dark matter?

Scientists realized they were missing something when they noticed that galaxies were spinning much faster than they should be, based on the gravitational pull of their visible matter. They were spinning so fast that they should have torn themselves apart. Something we can't see, which scientists have dubbed "dark matter," must be giving additional mass—and hence gravitational pull—to these galaxies.

Dark matter is thought to make up 27 percent of the contents of the universe. But it is not included in the Standard Model.

Accepted Science & Paradigms Which Are Likely Wrong

Scientists are looking for ways to study this mysterious matter and identify its building blocks. If scientists could show that dark matter interacts in some way with normal matter, "we still would need a new model, but it would mean that new model and the Standard Model are connected," says Andrea Albert, a researcher at the US Department of Energy's SLAC National Laboratory who studies dark matter, among other things, at the High-Altitude Water Cherenkov Observatory in Mexico. "That would be a huge game changer."

3. Why is there so much matter in the universe?

Whenever a particle of matter comes into being—for example, in a particle collision in the Large Hadron Collider or in the decay of another particle—normally its antimatter counterpart comes along for the ride. When equal matter and antimatter particles meet, they annihilate one another.

Accepted Science & Paradigms Which Are Likely Wrong

Scientists suppose that when the universe was formed in the Big Bang, matter and antimatter should have been produced in equal parts. However, some mechanism kept the matter and antimatter from their usual pattern of total destruction, and the universe around us is dominated by matter.

The Standard Model cannot explain the imbalance. Many different experiments are studying matter and antimatter in search of clues as to what tipped the scales.

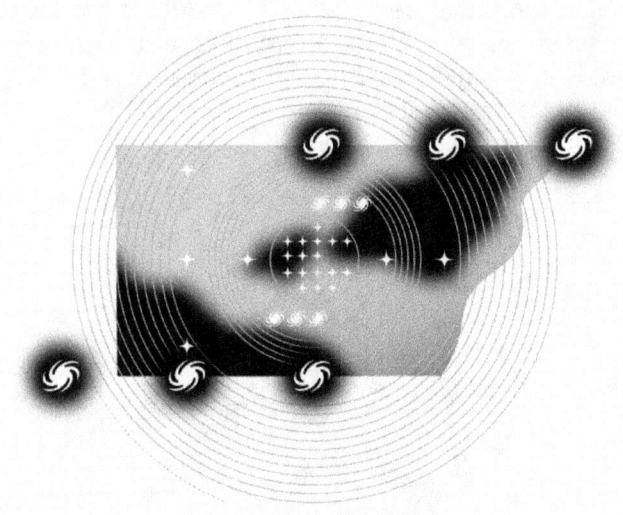

4. Why is the expansion of the universe accelerating?

Before scientists were able to measure the expansion of our universe, they guessed that it had started out quickly after the Big Bang and then, over time, had begun to slow. So it came as a shock that, not only was the universe's expansion not slowing down—it was actually speeding up. The latest measurements by the Hubble Space Telescope and the European Space Agency observatory Gaia indicate that galaxies are moving away from us at 45 miles per second. That speed multiplies for each additional megaparsec, a distance of 3.2 million light years, relative to our position.

This rate is believed to come from an unexplained property of space-time called dark energy, which is pushing the universe apart. It is thought to make up around 68 percent of the energy in the universe. "That is something very fundamental that nobody could have anticipated just by looking at the Standard Model," de Gouvêa says.

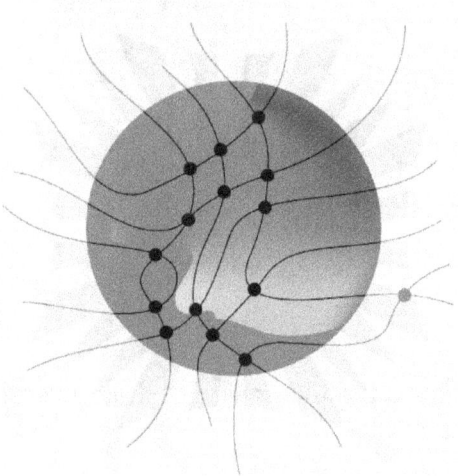

5. Is there a particle associated with the force of gravity?

The Standard Model was not designed to explain gravity. This fourth and weakest force of nature does not seem to have any impact on the subatomic interactions the Standard Model explains.

But theoretical physicists think a subatomic particle called a graviton might transmit gravity the same way particles called photons carry the electromagnetic force.

"After the existence of gravitational waves was confirmed by LIGO, we now ask: What is the smallest gravitational wave possible? This is pretty much like asking what a graviton is," says Alberto Güijosa, a professor at the Institute of Nuclear Sciences at UNAM.

More to explore

These five mysteries are the big questions of physics in the 21st century, Ramos says. Yet, there are even more fundamental enigmas, he says:

What is the source of space-time geometry? Where do particles get their spin? Why is the strong force so strong while the weak force is so weak? There's much left to explore, Güijosa says. "Even if we end up with a final and perfect theory of everything in our hands, we would still perform experiments in different situations in order to push its limits."

Accepted Science & Paradigms Which Are Likely Wrong

"It is a very classic example of the scientific method in action," Albert says. "With each answer come more questions; nothing is ever done."

10.0 Life Can't Exist Deep in the Earth

It was thought until just recently that life only existed on the surface of the Earth, the deep oceans, and in some caves. This is not the case and is a very limited view of the truth that life permeates the Earth to great depths.

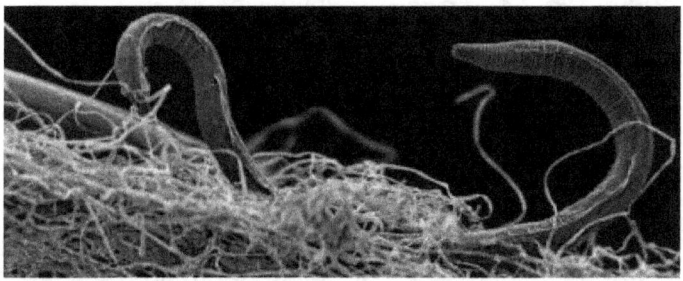

Scientists estimate that subterranean organisms constitute a massive amount of carbon, 245 to 385 times greater than that contained in all humans.

Organisms of Earth's deep underground constitute between 15 billion and 23 billion tons of carbon and occupy an estimated volume almost twice that of the oceans combined, scientists from the Deep Carbon Observatory reported yesterday (December 10) in advance of the annual meeting of the American Geophysical Union in Washington, DC.

The scientific team, which includes hundreds of researchers from all over the world, drilled boreholes kilometers below the continents and seafloor to sample microbes. The information collected by the scientists has allowed them to build models of the deep ecosystem and make the estimates of the deep life biomass.

The researchers found a stunning array of life, mostly microbial, and estimate that approximately 70 percent of the total number of Earth's bacteria and archaea organisms live in this realm. These microbes live at extremes of pressure, temperature, and nutrient and energy availability.

"Exploring the deep subsurface is akin to exploring the Amazon rainforest. There is life everywhere, and everywhere there's an awe-inspiring abundance of unexpected and unusual organisms," says Mitch Sogin, a scientist at the Marine Biological Laboratory and co-chair of the Deep Carbon Observatory's Deep Life team, in a statement.

Many questions remain as to how life spreads under the surface, which energy sources are the most important to sustain these organisms, and whether this is where life began on planet Earth.

The findings also suggest that extraterrestrial life may be similarly hidden underground.

"I think it's probably reasonable to assume that the subsurface of other planets and their moons are habitable, especially since we've seen here on Earth that organisms can function far away from sunlight using the energy provided directly from the rocks deep underground," Rick Colwell, a member of the Deep Carbon Observatory team from Oregon State University, tells *BBC News*.

Accepted Science & Paradigms Which Are Likely Wrong

Another Article about Deep Bioms:

Something odd is stirring in the depths of Canada's Kidd Mine. The zinc and copper mine, 350 miles northwest of Toronto, is the deepest spot ever explored on land and the reservoir of the oldest known water. And yet 7,900 feet below the surface, in perpetual darkness and in waters that have remained undisturbed for up to two billion years, the mine is teeming with life.

Many scientists had doubted that anything could live under such extreme conditions. But in July, a team led by University of Toronto geologist Barbara Sherwood Lollar reported that the mine's dark, deep water harbors a population of remarkable microbes.

The single-celled organisms don't need oxygen because they breathe sulfur compounds. Nor do they need sunlight. Instead, they live off chemicals in the surrounding rock — in particular, the glittery mineral pyrite, commonly known as fool's gold.

"It's a fascinating system where the organisms are literally eating fool's gold to survive," Sherwood Lollar said. "What we are finding is so exciting — like 'being a kid again' level exciting."

Sherwood Lollar is excited not only because of how peculiar the mine's rock-eating life seems, but also because of the growing realization that strange forms of life might not be so peculiar after all. Scientists are starting to find similar microbes in other deep spots, including boreholes, volcanic

Accepted Science & Paradigms Which Are Likely Wrong

vents on the bottom of the ocean and buried sediments far beneath the seafloor.

"The deep microbial realm reveals a biosphere that's more extensive, resilient, varied and strange than we had realized," said Robert Hazen, a mineralogist at the Carnegie Institution's Geophysical Laboratory in Washington, and co-founder of Deep Carbon Observatory, a global project to study the deep biosphere.

Cut off from light, air, and any connection to the surface, this shadowy realm seems more like an alien world than part of Earth. Hazen said exploring it could help us understand how life might have begun on other planets as well as on our own. We might even find alien-like creatures living undetected right beneath our feet.

Lots of life at the bottom

Sherwood Lollar's work builds on a 2018 report by Deep Carbon Observatory scientists who tried to map the total extent of Earth's deep biosphere comprehensively for the first time.

In the eye-opening report, a team led by Cara Magnabosco, a geobiologist at the Swiss technical university ETH Zurich, estimated that some 5×10^{29} cells live in the deep Earth: that's five-hundred-thousand-trillion-trillion cells. Collectively, they weigh 300 times as much as all living people combined. The team describes this hidden ecosystem as an "underground Galapagos."

Accepted Science & Paradigms Which Are Likely Wrong

An exterior shot of Kidd Mine.

The denizens of the deep are an exotic bunch even beyond their appetite for solid rock. One species, the microbe *Geogemma barossii,* can live at temperatures of 250 degrees Fahrenheit — well above the boiling point of water and close to the theoretical limit at which vital organic molecules start to disintegrate.

Separate studies of material drilled near the Mariana Trench in the Pacific Ocean hint that some organisms could be living six miles below the seafloor, limited only by the heat at such tremendous depths. Laboratory experiments show that some microbes can tolerate pressures 20,000 times higher than the air pressure at sea level, meaning that there are almost certainly more extreme ecosystems out there than the one in the Kidd Mine.

"We're finding that we don't really understand the limits to life," Sherwood Lollar said.

Accepted Science & Paradigms Which Are Likely Wrong

The pace of existence in the deep also seems radically different from that on the surface. In ancient environments like the trapped waters at the bottom of the Kidd Mine, food and energy are scarce. To compensate, cellular metabolism slows almost to a standstill.

"Many of the microbes may survive for thousands of years or more without dividing, just replacing their broken parts," said Karen Lloyd, a University of Tennessee microbiologist who studies life at the bottom of the ocean.

There are so many deep microbes that, despite a seemingly lazy existence, they collectively exert a huge impact on their habitats. For instance, a community of cells on the ocean floor consume methane gas that bubbles up from ancient sediment. "Deep subsurface microbes eat massive amounts of methane that would otherwise be released," Lloyd said, helping curb atmospheric levels of a potent greenhouse gas.

Back to beginnings

One of the big questions facing Sherwood Lollar is how the deep-life community at the Kidd Mine is related to those found in other mines or stretched out beneath the oceans. "The number of systems we've looked at so far really is limited," she said, "but they probably had a single origin at some point in life's 4-billion-year history."

If so, there should still be clues about when and how life first colonized the deep.

Accepted Science & Paradigms Which Are Likely Wrong

Fossils show that surface life has changed enormously over billions of years, but slow-motion deep life may retain much of its primitive characteristics. That's especially true at the Kidd Mine, which is in one of the oldest, most stable portions of Earth's crust. (The rock in and around the mine have lain undisturbed for 2.7 billion years, and have been cool enough to support life for at least 2 billion years.)

Cara Magnabosco and colleagues collect ancient water samples 4,300 feet deep within the Beatrix Gold Mine, South Africa to investigate the diversity and abundance of deep microbes.

Sherwood Lollar wants to sequence the genes of the Kidd Creek microbes and then do a 23andMe-style analysis to unravel their kinship to other residents of the deep Earth:

Accepted Science & Paradigms Which Are Likely Wrong

Are they all still close relatives, or have they diversified and adapted significantly to their local environments? It's a delicate project, but she hopes to have results within a year or two.

Such studies could offer hints about where life first arose on Earth. Charles Darwin imagined the beginning might have occurred in a warm little pond, but "there's absolutely no reason why it could not have been a warm little rock fracture," Sherwood Lollar said. In many ways, she noted, sulfur-breathing microbes living beneath thick, protective layers of rock would have been well suited to the brutal conditions on our planet when it was young.

Another, even wilder possibility is that life originated more than once, with other forms still surviving somewhere on Earth. "We've literally only scratched the surface of the deep biosphere," Hazen said. "Might there be entire domains that are not dependent on the DNA, RNA and protein basis of life as we know it?" Perhaps we just haven't found them yet.

Paul Davies, a physicist at Arizona State University, has long advocated systematic searches for such "shadow life." The recent forays into the deep biosphere show how it might be done. Since known organisms cannot survive above 250 degrees Fahrenheit, Davies suggests going to extreme environments (around undersea volcanic vents, for instance) and checking for anything that appears alive at temperatures around 300 to 400 degrees Fahrenheit.

Accepted Science & Paradigms Which Are Likely Wrong

"That would stand out as a candidate for shadow life," he said.

Ever cautious, Sherwood Lollar points out that she hasn't found any evidence of shadow life at the Kidd Mine. But she heartily agrees that scientists need to keep a wide-open mind about what could be lurking within the deep world: "We see only what we look for. If we don't look for something, we miss it."

Accepted Science & Paradigms Which Are Likely Wrong

11.0   Can People Time Travel?

After writing several books on Time Travel I think our scientists still have a deep misunderstanding of time.

I did lots of research on individual Time Travel experiences for my book titled "Real Time Travel Stories From a Psychic Engineer".

This led me to some fascinating stories which shed some light upon the structure of time and that some type of time portals do exist.

Here are a couple of those stories:

Bold Street, Liverpool, England

Bold Street is a site with the most experiences of time slips so far discovered. The below stories indicate that there is some type of time warp on this street which some people pass through.

Accepted Science & Paradigms Which Are Likely Wrong

The Liverpool Time Slips and Mysterious Occurrences in Bold Street are numerous. This location seems to be some type of time portal. Several stories follow.

The subject of time has always intrigued us. Is it as set as we have always believed? Or does time loop back on itself, giving us a glimpse of a shadowy past out of the corner of our eye.

Was is just our imagination that made us believe we had seen an object or building change before our very eyes, and seem as though we had stepped back into the past?
When this happens we usually shake our heads and put it down to imagination.

But over the last few decades, something strange has been happening in or near Bold Street, Liverpool England. Not just a glimpse of the past, but full immersion into the strange and mysterious world of English History, if only for a few moments at a time.

The strange thing about the Bold street time slips is the actual time and place they are set. In the following cases, the people involved do not go back really far, but seem to visit a particular decade or decades.

So far, most of the sightings have centered around the 1950s and '60s. This is strange in itself. Most time travel experiences seem to take the recipient back to the 18th or 19th century. But not in this case.

Accepted Science & Paradigms Which Are Likely Wrong

Are these people simply copying each other in their experiences, or are they genuinely taking a step back in time?

The answer to this has to take into account whether they are doing it deliberately to get noticed. In other words are the perpetuating a hoax?

Another explanation could be mass hallucination. And last but not least, they really are experiencing this strange phenomena!

The most important point is, the very first person that had this experience, obviously totally believed in what he saw, heard and felt.

So, does time flow like a river? Or does it twist and turn, going forward then sweeping back, picking up historic events and placing them down in front of you, if only for a few moments?

Accepted Science & Paradigms Which Are Likely Wrong

In this first tale, we find Frank and his wife out for a stroll in Liverpool town center. It is 1996.

His wife decided that she wanted to go and buy a book at Waterstone's the large book store, and they started to walk towards the area of the shop.

As they approached Bold Street, Frank decided to go to another shop first, but bumped into his friend, and stopped to chat in the street. His wife went ahead without him.

A few moments later, Frank said goodbye, visited his shop and turned to go back to meet his wife. After reaching Bold Street, he headed on towards the bookstore. As he approached, he glanced up and was surprised to see the name, Cripps above the door. As he was about to cross over to see what was going on, a van swept past him with the name Cardin's on the side. The van driver honked his old fashioned horn and drove past.

Looking around, Frank suddenly realized that things were not quite what they should be. He looked at the cars driving past and realized that they were all old fashioned vehicles such as people would drive back in the 50's and 60's.

And then he noticed the people. Men were wearing hats and macs, and the women were dressed in head scarves, full skirts and had old fashioned hair styles such as women wore just after the war.

By this time, Frank was beginning to feel slightly freaked out. He carried on crossing the road and headed towards the store.

Accepted Science & Paradigms Which Are Likely Wrong

As he got closer he noticed in the window there were handbags, shoes, and umbrellas. Suddenly he saw a young woman looking up at the shop sign. She looked confused.

She was wearing modern clothes and as she saw him approaching, she smiled at him.

Frank went into the shop, closely followed by the young woman. When they entered he was surprised and pleased to see that it had indeed turned back into a bookshop. The young woman smiled, shook her head and said, 'that was strange, I thought it was a new clothes shop!' then she walked away looking extremely puzzled.

This may sound an unlikely tale, but the odd thing about it is that Frank was, in fact, a former Police officer who was used to dealing in facts, and definitely wasn't the type of person who would believe in the paranormal.

Accepted Science & Paradigms Which Are Likely Wrong

Frank never stopped talking about it. Was this a time slip? Evidently Cripps was a women's shop that sold clothes and other goods decades before!

And Cardin's was also a well-known Liverpool firm that owned vans around the same time.

The second story concerns a young girl by the name of Imogen. She had decided to go into Liverpool to buy her sister Abigail a few things for her new baby. Upon arriving she was happy to see a new MotherCare store that had opened up on the corner of Lord Street and Whitechapel.

She wandered around the store, and picked up a few baby items such as cardigans, baby bibs, and gloves. She was surprised to see how cheap the items were, but thought they were on offer as the store had just opened. Taking them to the counter, she tried to pay with her credit card. The staff member looked at her suspiciously, and went off to get the manager.

When she came back, she looked at the card and told Imogen that they didn't take cards. So, disappointed, Imogen went and put the items back as she hadn't any money with her.

When she got home, she told her mother what had happened. Her mother was surprised and really puzzled. 'That store closed years ago,' she said. 'There is a bank there now, in fact that's where I have my account'.
Not believing her, Imogen took her mother back to the same place the next day. Sure enough the

store wasn't there. It was a bank, just as her mother had told her.

The third tale is of a young man named Sean, who, while shoplifting in Liverpool back in 2006, ran away from a Security Guard and headed down Hanover Street. Trying to shake off the Guard, Sean, 19, turned into a dead end street called Brookes Alley.

By this time he was out of breath and started to get a tight sensation in his chest. He soon realized that actually it wasn't a problem with him, but the atmosphere around him.

He waited for the Guard to come around the corner after him, but he never appeared. So, thinking he had given him the slip, he sauntered back out and started to walk down Hanover Street again. But he soon realized that something was wrong.

The road looked different, and so did the pavement. He noticed cars driving by that looked very old fashioned, and the road works that he knew were there, were now gone.
Soon he saw that the people around him were wearing strange clothes. Crossing over to Bold Street, he noticed that there were traffic lights where they weren't before, and bushes growing around the Lyceum, near a bar that he recognized.

He carried on walking. Soon he began to feel that something was not quite right. Then he began to panic. He realized that somehow he had stepped back in time. And the time slip was not going away.

Accepted Science & Paradigms Which Are Likely Wrong

Then he remember his Cell phone. Taking it out of his pocket, he tried to get a signal, but of course it didn't work. Eventually he began to really panic, but soon spotted a kiosk selling newspapers and headed over.

Leaning over the Stand, he took a look at the front page of the Daily Post. There in bold lettering was the date. 18th May 1967.

He wondered what to do. What happens if he can't get back to his own time? What about family and friends?
So, speeding up his pace, he reached H. Samuel the Jewelers, and tried his phone once again. This time it worked. Sighing with relief he looked around and realized that he had returned to the present. But the strange thing was, he could still see, down the end of the road, people still walking around in 1967.

By this time Sean had seen enough, and dived onto a bus to go home. When he was interviewed by the local newspaper later, he stated over four times, the exact account.

Now, you may think that Sean was making the story up to escape from the guard. But the strange tale didn't end there. When the Security Guard was interviewed, he stated that when he ran after Sean, and turned down the dead end alley after him, he said that Sean had completely disappeared!

When the newspaper checked out the facts of Sean's story, they found that everything he said was historically accurate.

Accepted Science & Paradigms Which Are Likely Wrong

These three stories are just the tip of the iceberg. There are many tales from around Liverpool that tell of time slips, ghosts and other strange phenomenon. The stories keep coming thick and fast, and of course the more tales, the more likely people will start to believe that they are all being made up, or as the saying goes, Urban Tales. So, what do you think? Real life time slips, imagination, mass hallucination or purely tales that have started out as fun but have turned into the greatest Urban Legends of all time.

My opinion is that, yes, something did happen.

Probably to the first guy, Frank who was just out shopping with his wife. The others? Maybe it was a case of mistaken roads, taking a wrong turning or just a glitch in the person's memory. By the time they get home they totally believe what happened.

Accepted Science & Paradigms Which Are Likely Wrong

Or is it true? There are so many cases concerning Bold Street, and just about anywhere else in Liverpool, that maybe, just maybe they are all living on top of the biggest time slip phenomena in the World.

Accepted Science & Paradigms Which Are Likely Wrong

Sir Victor Goddard

A Flight through Time

Sir Victor Goddard's trip into the unexplained involved an airplane flight. This was a much more personally harrowing experience.

In 1935, while a Wing Commander, Goddard flew a Hawker Hart biplane to Edinburgh, Scotland, from his home base in Andover, England, for a weekend visit. On the Sunday before flying back, Goddard visited an abandoned airfield in Drem, near Edinburgh, this location being closer to his final destination than the airport at which he landed. The Drem airfield, constructed during the First World War, was a shambles. The tarmac and four hangars were in disrepair, barbed wire divided the field into numerous pastures, and cattle grazed everywhere. It was now a farm, and completely useless as an airfield.

On Monday, Goddard began the flight back to his home base. The weather was dark and ominous, with low clouds and heavy rain. Goddard was flying in an open cockpit over mountainous terrain without radio navigational aids or cloud flying instruments. Rain began beating down on his forehead and onto his flying goggles badly which obscured his vision. He thought he could climb above the clouds, but he was wrong. He made it to 8,000 feet, looking for a break in the clouds. There was none.

Suddenly Goddard lost control of his plane. It began to spiral downward. He struggled with the controls. He could speed up or slow down, but he could not stop the spin. He was unsure of his location, but knew he was falling rapidly and might smash into the mountains before coming out of the clouds. The sky became darker, the clouds turning a strange yellowish-brown. The rain came down even more heavily. Goddard's altimeter showed he was only a thousand feet above the ground and dropping rapidly. At two hundred feet and still spiraling downward, he began to see a bit of daylight through the murky gloom, but his spiral toward seemingly inevitable death was far from over.

Goddard was now flying at 150 miles per hour. He emerged from the clouds over "rotating water" that he recognized as the Firth of Forth. He was still falling.

Accepted Science & Paradigms Which Are Likely Wrong

Suddenly, he saw directly before him a stone sea wall with a path, a road, and railings on top of it. The road seemed to be slowly rotating from left to right. The cloud cover was down to forty feet. Goddard was now flying below twenty feet and was within an instant of tragedy. A young girl with a baby carriage ran through the pouring rain. She ducked her head just in time to avoid Hart's wingtip. Goddard succeeded in leveling out his plane after that. He barely missed striking the water after clearing the sea wall by a few feet.

He was now flying only several feet above a stony beach. Fog and rain obscured all distant visibility, but Goddard somehow located his position. He identified the road to Edinburgh and soon was able to discern, through the gloom, the black silhouettes of the Drem Airfield hangars ahead of him, the same airfield he had visited the day before. The rain became a deluge, the sky grew even darker, and Goddard's plane was shaken violently by the turbulent weather as it sped toward the Drem hangars-and into a different world.

Suddenly, the sky turned bright with golden sunlight. The rain and the farm had vanished. The hangars and the tarmac appeared to have somehow been rebuilt in a brand-new condition. There were four planes lined at the end of the tarmac. Three were standard Avro 504N trainer biplanes; the fourth was a monoplane of an unknown type-the RAF had no monoplanes in 1935. All four airplanes were bright yellow. No RAF airplanes were painted yellow in 1935. The airplane mechanics were wearing blue overalls. RAF mechanics never wore anything but brown overalls when working in hangars in 1935.

It took Goddard only an instant to fly over the airfield. He was only a few feet above the ground- just high enough to clear the hangars-but apparently none of the mechanics saw him or even heard his plane. As he sped away from the airfield, he was again engulfed by the storm. He forced his plane upward, flying at 17,000 feet and then, for a time, at 21,000 feet. He managed to return to his home base safely.

Goddard felt elated when he landed. He then made the mistake of telling fellow officers about his eerie experience. They looked at him as if he were drunk or crazy. Goddard decided to keep silent about what had happened to him. He did not want a discharge from the RAF on mental grounds.

In 1939, Goddard watched as RAF trainers began to be painted yellow and the mechanics switched to blue coveralls. The RAF introduced a new training monoplane exactly like the one he had seen in his flight over Drem. It was called the Magister. He learned that the airfield at Drem had been refurbished.

Accepted Science & Paradigms Which Are Likely Wrong

Another twenty-seven years went by, but Goddard never forgot what had happened. He played it through over and over in his mind. It was not until 1966 that he wrote of this experience. Over the years he had become convinced that there was no way he could have known that the RAF would change the colors of their trainers and their mechanics' overalls four years before these changes took place. Goddard finally concluded that he must have glimpsed the future-or even traveled into it-for a brief moment in time.

Accepted Science & Paradigms Which Are Likely Wrong

12.0 Civilization On Earth is Recent

I'm frankly amazed that archeologists and anthropologists can ignore a large amount of evidence of intelligent life existing on Earth before home sapiens.

Also, "accepted" archeologists have a much shorter timeline for massive constructions built in pre-history than the evidence suggests.

This is a good example of closed mindedness and supposedly objective scientists sticking to outdated ideas to support their theories of the past.

In this chapter we will look at examples of several things:

A) Large constructions which were built well before accepted timeframes that civilization existed.
B) Evidence that Giants lived on the Earth in human history and the distant past
C) That some type of civilization existed on Earth over 300 million years ago.

12.1 Large Constructions in the Deep Past

Most Archeologists would say that man didn't start building large structures until the age of the Pyramids several thousand years B.C.

Or maybe include Stonehenge which was built about 3,000 B.C.

But anything a lot older just doesn't exist. Well I'm sorry to break their bubble, but there are lots of structures with verified ages which are much older and indicate that some previous civilizations existed well before scientists say that civilization existed.

Gunung Padang, Western Java 22,000 B.C.

(How the Pyramid may have looked when it was built)

Gunung Padang is a megalithic site located in Karyamukti village, Cianjur regency, West Java Province of Indonesia, 30 kilometers (19 mi) southwest of the city of Cianjur or 6 kilometers (3.7 mi) from Lampegan station. It has been called the largest megalithic site in all of Southeastern Asia, and has produced controversial carbon dating results which, if confirmed, would suggest that construction began as far back as 22,000 B.C.

Located at 885 meters (2,904 ft) above sea level, the site covers a hill in a series of terraces bordered by retaining walls of stone that are accessed by about 400 successive andesite steps rising about

95 meters (312 ft). It is covered with massive rectangular stones of volcanic origin. The Sundanese people consider the site sacred and believe it was the result of King Siliwangi's attempt to build a palace in one night. The asymmetric step pyramid faces northwest, to Mount Gede and was constructed for the purpose of worship.

A picture view of the top of the Pyramid:

Another view of stones implanted on top of the pyramid:

Accepted Science & Paradigms Which Are Likely Wrong

Another picture of this pyramid from another perspective. It really looks like a stepped pyramid except in dirt on the outside, with stone at many places inside:

There also appears to be an ancient cement which is a mixture of clay, iron, and silica. The cement was used as an adhesive to glue the rocks together, containing up to 45% iron:

Accepted Science & Paradigms Which Are Likely Wrong

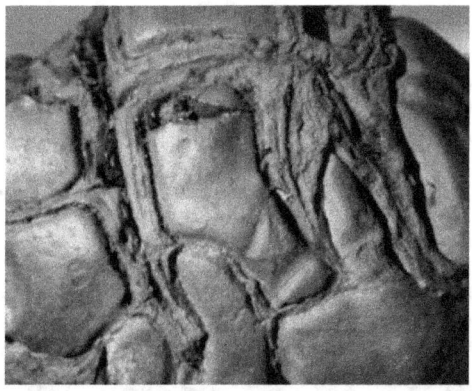

Some Of The Layers Between 5 – 12 Meters Deep Are Confirmed To Be Between 14,500 B.C. -25,000 B.C.

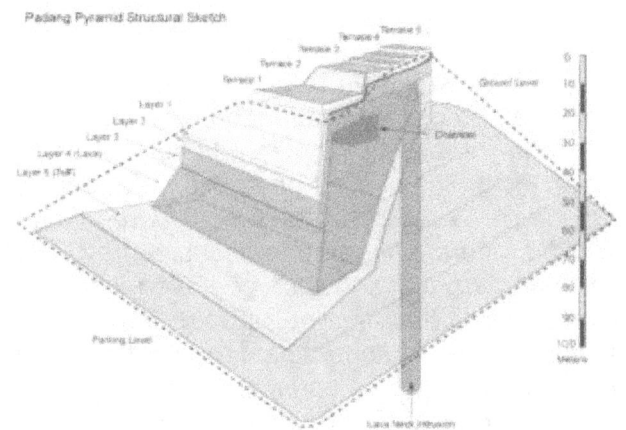

This suggests that the civilization that built the original structure had their pyramid built upon over the millennia. Because of the vast amounts of time that elapsed between each development of the pyramid, the peoples that built on top of it probably had no idea what lay beneath them!

Accepted Science & Paradigms Which Are Likely Wrong

Bosnian Pyramids, Bosnia 30,000 B.C.

Dr. Sam Semir Osmanagich, Ph.D. has the following story about the Bosnian Pyramids:

In April 2005 I first traveled to the town of Visoko, 20 miles northwest from Sarajevo, the capital of Bosnia-Herzegovina. My attention was caught by two regularly shaped hills, which I later named the Bosnian Pyramids of the Sun and the Moon. For thousands of years locals have considered those hills to be natural phenomena because they were covered by 3-foot of soil and vegetation. However, when I first saw their triangular faces, same slopes, obvious corners and orientation toward the cardinal points, I knew that they had to be constructed by a force other than nature. Since I had been investigating pyramids for decades I knew that the pyramids found in China, Mexico, Belize, Guatemala or El Salvador presented the same case of pyramids covered by dirt and vegetation.

Excavations have been carried out for a number of years now and have found some really interesting buried slabs which make the whole structure look a lot more like an intelligent construction. Here are some of the pictures of what was found—flat stones looking like a tiled floor:

Another set of squarish slabs on the pyramid:

Accepted Science & Paradigms Which Are Likely Wrong

Analysis of the slabs also shows them to be some type of early concrete. Concrete was purportedly invented by the ancient Romans. If it was developed much earlier it would indicate some high building technology in pre-history.

Other pyramids and ancient tunnels in this same complex have also been found. Here is one of the tunnels:

There are actually three pyramids in this area. Here is a diagram of how all three of the pyramids are laid out:

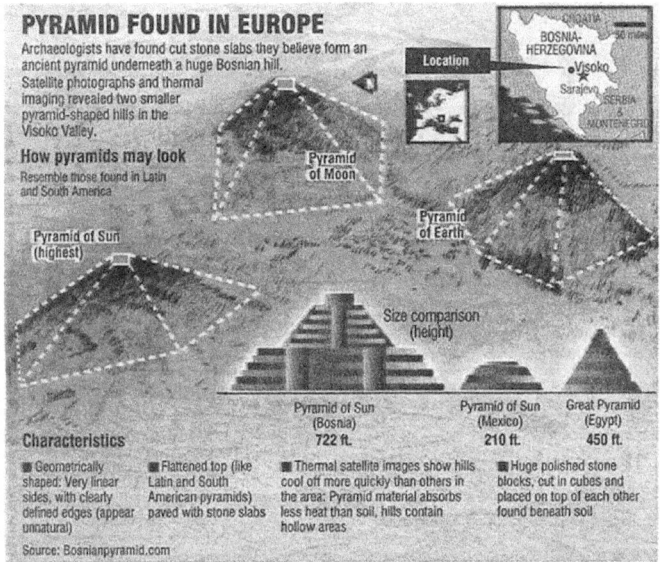

For these pyramids to be so covered by dirt and vegetation and having many flat slabs below the dirt, it must have been constructed and later abandoned many thousands of years ago.

… Accepted Science & Paradigms Which Are Likely Wrong

12.2 Giants in the Earth

There are many claims that Giants existed on the Earth in historical and even recent times. The main evidence is based on skeletons found of people who were seven to eight feet taller or much taller. They also had two rows of teeth and six fingers on their hands.

My book titled "The History of Antediluvian Giants" covers many of these records.

There are many records and skeletons which have been found in excavations of an ancient race of giants. These giants lived simultaneously with Homo sapiens and might have even been a close branch of the tree of humanity. In those civilizations where Giants may have been involved we will identify them. (Many also believe that in the nineteenth and early 20th centuries the Smithsonian Museum destroyed many of these giant skeletons.)

Many of those ancient societies may have been started by Giants or had Giants involved as leaders of those civilizations. A good example is King Og who was a real historical personage:

Accepted Science & Paradigms Which Are Likely Wrong

In the Bible read The Book of Numbers, Chapter 21, and Deuteronomy, Chapter 3, which has an account of him:

"Next we turned and headed for the land of Bashan, where King Og and his entire army attacked us at Edrei. But the Lord told me, 'Do not be afraid of him, for I have given you victory over Og and his entire army, and I will give you all his land. Treat him just as you treated King Sihon of the Amorites, who ruled in Heshbon.'

"So the Lord our God handed King OG and all his people over to us, and we killed them all. Not a single person survived. We conquered all sixty of his towns—the entire Argob region in his kingdom of Bashan. Not a single town escaped our conquest. These towns were all fortified with high walls and barred gates. We also took many unwalled villages at the same time. We completely destroyed the kingdom of Bashan, just as we had destroyed King Sihon of Heshbon. We destroyed all the people in every town we conquered—men, women, and children alike. But we kept all the livestock for ourselves and took plunder from all the towns.

"So we took the land of the two Amorite kings east of the Jordan River—all the way from the Arnon Gorge to Mount Hermon. (Mount Hermon is called Sirion by the Sidonians, and the Amorites call it Senir.) We had now conquered all the cities on the plateau and all Gilead and Bashan, as far as the towns of Salecah and Edrei, which were part of Og's kingdom in Bashan. **(King Og of Bashan was the last survivor of the giant Rephaites. His bed**

was made of iron and was more than thirteen feet long and six feet wide. It can still be seen in the Ammonite city of Rabbah.)"

*Og's destruction is told in Psalms 135:11 and 136:20 as one of many great victories for the nation of Israel, and the Book of Amos 2:9 may refer to **Og as "the Amorite" whose height was like the height of the cedars and whose strength was like that of the oaks.***

Many Giants found in North America were eight to nine feet tall, had reddish hair, six fingers on each hand, and six toes on each foot. They also had two rows of teeth.

Many other Giants are purported to have lived in ancient civilizations and we will note them as we learn more about these age old societies and cities. Below are some charts showing some of the sizes of giant skeletons found around the world:

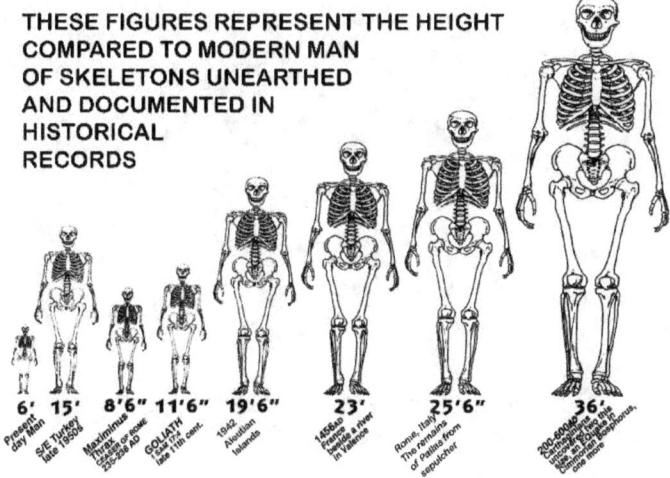

THESE FIGURES REPRESENT THE HEIGHT COMPARED TO MODERN MAN OF SKELETONS UNEARTHED AND DOCUMENTED IN HISTORICAL RECORDS

And here is a map of Giant skeletons found just in the United States:

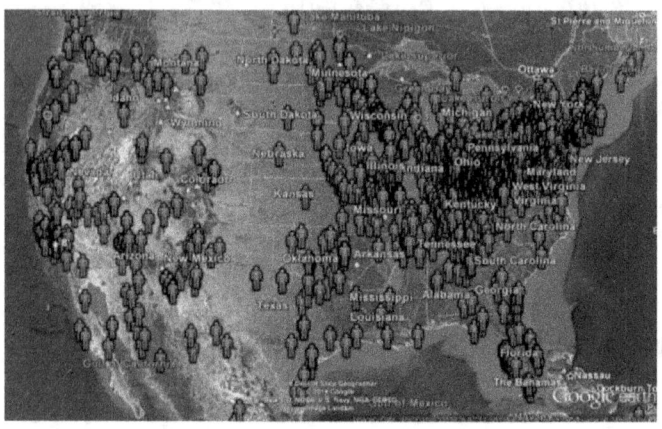

Another chart shows the measurements of many giant skeletons found in the USA and has a key of where they were found.

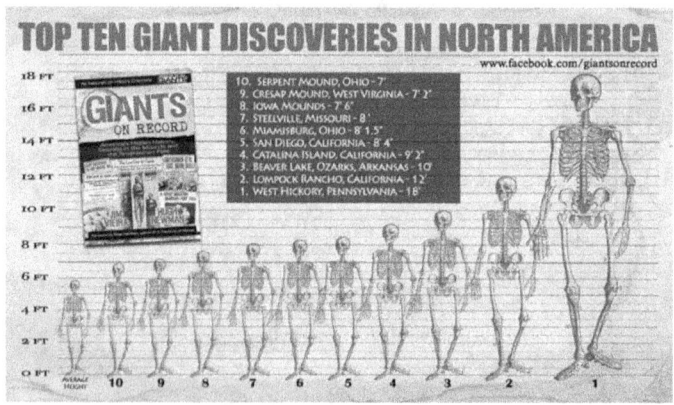

You can see that evidence of Giants found is a worldwide phenomenon.

12.3 Civilization 300 Million Years Ago

Out of Place Artifacts are a fascination of mine. These are intelligently made objects which come from rock strata or coal mines so they have to be millions of years old. (I've got several books on this topic and my most comprehensive is titled "The Encyclopedia of out of Place Artifacts").

They come from times when Man obviously did not exist. These objects used to be dismissed by serious investigators but they keep popping up to the point that we can make some theories about them.

One of my theories is the subject of my book titled "A 300 Million Year Old Civilization Existed on Earth".

There are a number of these objects which have detailed descriptions and pictures in the book. Here we show a quick list of artifacts from that era and one of them described in detail below:

Objects from 300 Million Years Ago

- An Iron Face Mask-300 Million Years Old
- Gold Thread-320 to 360 Million Years Old
- Ancient Screw-300 to 320 Million Years Old
- Iron Pot from Oklahoma-312 Million Years Old
- Mysterious Bell-298 to 320 Million Years Old
- A Screw Object Hidden Inside Stone 300 Million Years Old
- Ancient Modern Tools-300 Million Years Old
- Mechanical Part-300 Million Years Old
- Wheel in Coal Mine- 300 Million Years Old
- France-Tools in Rock 300 Million Years Old

Accepted Science & Paradigms Which Are Likely Wrong

An Iron Mask of a Face-300 Million Years Old:

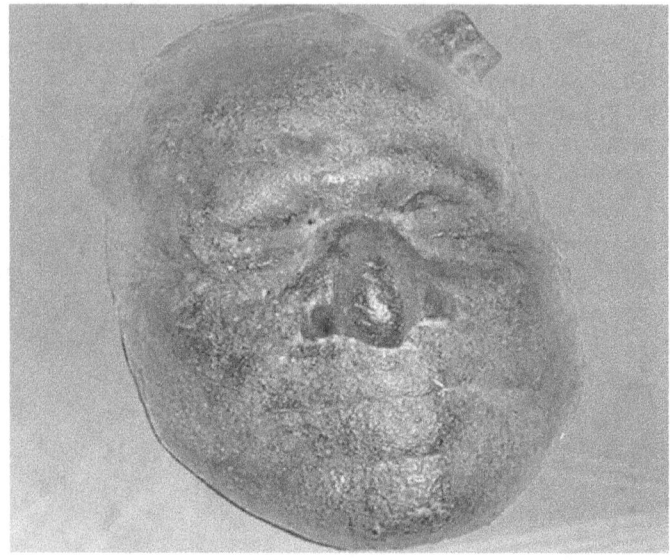

It is a once smelted, solid iron recreation of a face, whose owner could have lived an unimaginably long time ago.

Within the claim written by John D Morris PHD, quote:

"I was recently contacted by an older lady, who grew up in the coal mining area of Appalachia.

Her ancestors having lived in the area for generations, her now deceased, was a miner who had once made a remarkable discovery embedded within a coal seam—a human face made from cast iron!

Accepted Science & Paradigms Which Are Likely Wrong

Like most people, they had been taught that coal is far too old to contain any human artifacts. The miner was so proud and perplexed by his find, it eventually became family heirloom and was simply named "Man".

As a large, heavy object, it was eventually used as an ornament, decades later becoming stored among his belongings. She distinctly remembers her father's story of its discovery and the care he had taken with this prized object, having recently rediscovered it among her father's possessions."

Accepted Science & Paradigms Which Are Likely Wrong

13.0 That Mars Travel Takes Months or Years

Humans Could Reach Mars In One Month with a 200,000 Kmph Nuclear Rocket.

Planning to move to Mars when it's feasible? Worry not, a rocket company is currently testing nuclear rockets that will cut down travel time to Mars to just one month. Current rockets take about seven months to reach Mars after covering 480 million kilometers (300 million miles).

Every journey in space is extremely dangerous. Longer journeys have higher scope of mechanical failures and encountering other space hazards. The pilot step towards minimizing risks to human lives is by minimizing the travel time and that's exactly what this company is trying to do.

Called "Ad Astra Rocket Company", this Costra Rica-based rocket maker completed a record 88-hour long test of its Vasimr VX-200SS plasma rocket. Conducted at a facility near Houston in Texas, the test set a world record in terms of high-power endurance in electric propulsion.

Accepted Science & Paradigms Which Are Likely Wrong

The Vasimr rocket (Variable Specific Impulse Magnetoplasma Rocket) uses an engine based on nuclear reactors to heat plasma to over two million degrees. Upon attaining this temperature, hot gas is expelled out of the engine's rear through magnetic fields, causing the rocket to move forward. With this, the rocket can achieve speeds as high as 197,950 kilometers/hour (123,000 mph).

With this astounding feat, Ad Astra not only hopes to make travel time in space shorter, but also safer for all parties involved. Current chemical rockets take much longer and open up windows of failures that could plague space travelers and astronauts. With growing interest in space tourism and exploration, it could be normal to be a spacefarer for everyone within this century with SpaceX, Blue Origin and Virgin Galactic hoping to make space travel the norm.

Accepted Science & Paradigms Which Are Likely Wrong

14.0   How Long Can We Live?

The longest lived person generally acknowledged was Jean Calumet of France who live to 122 years old. Most scientists think we can't live much beyond 100 years old.

In 2008 I decided to answer the question for myself of how long people can live. I did months of research on the topic and found hundreds of people who had lived well beyond one hundred years, many to their mid one hundreds, and some over 200 years.

This led to my book "Physical Immortality: A History and How to Guide" where I did my best to understand how people can live that long.

Here is a small sample of some of the long lived person's records I collected:

Ages 150-159

Christian Jacobsen Drakenberg died at 150 years in 1772. A sailor for 91 years, he fought in the war against the Swedes, then became a merchant seaman. In 1694, he was taken prisoner by Algerian pirates but set free after 15 years of slavery, he resumed his life as a seaman. In 1737, at the age of 110, he married a widow of 60 years. He was known as 'the old man of the north'.

Even in old age Drakenberg was bursting with strength. Whoever would shake his hand, never forgot the experience and ventured no second attempt. It was reported that after death his body mummified and did not rot. (Similar to reports on Yogananda)

In the chancel of the Honigton Church, Wiltshire, is a black marble monument to the memory of G. Stanley, a gentleman, who died in 1719, aged one hundred and fifty-one.

And in Acsadi & Nemeskeri, p.17 & Toronto Evening Telegram, 9 Sept., 1939; 26 April, 1942.

Thomas Parr, 152, died 1635, in England. Thomas Parr (or Parre), among Englishmen known as "old Parr," was a poor farmer's servant, born in 1483. He remained single until eighty. His first wife lived thirty-two years, and eight years after her death, at the age of one hundred and twenty, he married again. Until his one hundred and thirtieth year he performed his ordinary duties, and at this age was even accustomed to thresh.

Accepted Science & Paradigms Which Are Likely Wrong

LI CHING-YUN: The Longest Lived person of record-256 Years

Below is an excerpt of an article from the New York Times:

The New York Times, Saturday, May 6, 1933

LI CHING-YUN DEAD; GAVE HIS AGE AS 197

"Keep Quiet heart, Sit Like a Tortoise, Sleep Like a Dog," His advice for a Long Life. Inquiry Put Age At 256.

Reported to have buried 23 wives and had 180 descendents – sold herbs for first 100 years. Peiping, May 5 – Li Ching-Yun, a resident of Kaihsien, in the Province of Szechwan, who contended that he was one of the world's oldest men and said he was born in 1736 – which would make him 197 years old – died today.

A Chinese dispatch from Chungking telling of Mr.

Accepted Science & Paradigms Which Are Likely Wrong

Li's death said he attributed his longevity to peace of mind and that it was his belief every one could live at least a century by attaining inward calm.

Compared with estimates of Li Ching-Yun's age in previous reports from China, the above dispatch is conservative. In 1930 it was said Professor Wu Chung-chien, dean of the department of Education in Minkuo University, had found records showing Li was born in 1677 and that Imperial Chinese Government congratulated him on his 150th and 200th birthdays.

A correspondent of The New York Times wrote in 1928 that many of the oldest men in Li's neighborhood asserted their grandfathers knew him as boys and that he was then a grown man.

According to the generally accepted tales told in his province. Li was able to read and write as a child, and by his tenth birthday had traveled in Kansu, Shansi, Tibet, Annam, Siam and Manchuria gathering herbs. For the first hundred years he continued at this occupation. Then he switched to selling herbs gathered by others.

Wu Pei-fu, the warlord, took Li into his house to learn the secret of living to 250. Another pupil said Li told him to "keep a quiet heart, sit like a tortoise, walk sprightly like a pigeon and sleep like a dog."

According to one version of Li's married life he had buried away twenty-three wives and was living with his twenty-fourth, a woman of '60.' Another account, which in 1928 credited him with 180 living descendants, comprising eleven generations, recorded only fourteen marriages. This second authority said his eyesight was good; also, that the finger nails of his right hand were very long, and "long" for a Chinese might mean longer than any finger nails ever dreamed of in the United States.

One statement of The Times correspondent which probably caused skeptical readers to believe Li was born more recently that 1677, was that "many who have seen him recently declare that his facial appearance is no different from that of persons two centuries his junior."

Several years of additional research and thinking resulted in my book "The 10 Principles of Personal Longevity".

I strongly believe that by following the ten principles each individual can learn to live decades longer.

Here are the 10 Principles laid out in order:

Accepted Science & Paradigms Which Are Likely Wrong

The 10 Principles of Personal Longevity

1) Real Long Lived Persons Exist
People really have lived a long time-so you can do it too
2) Define Your Purpose in Life
Know your life purpose-To live life with meaning
3) Enable Your Life Urge
Know without doubt that you will live a long and happy life
4) The Importance of a Spiritual Connection
A spiritual connection is important for happiness & long term health
5) Having Love in your Heart
Unconditional Love is is real-It will make you happier and healthier
6) Activate your Vital Forces
Improve the vitality of your energy body for health and to enjoy life more
7) The Science of Longevity
Use new therapies and discoveries from Science & Medicine
8) Keep your Physical Body Healthy
Eat a proper diet, use herbal supplements, and exercise
9) Use Your Intuition for Safety
Learn to use your intuition to keep you safe
10) Implement the above principles in your life
Implement these principles for long term health, greater happiness, and extended longevity

Accepted Science & Paradigms Which Are Likely Wrong

15.0 Cryptids (Like Bigfoot) Really Do Exist

Cryptids are legendary animals who have been reported many times but whom science has yet to prove that they exist. (See my book "Bigfoot Mysteries and Some Answers").

There are thousands of stories of persons who have seen them. Here is just one sighting story out of many about Bigfoot, and then the map of reported sightings.

The Ouachita Mountains

Kathy Strain, a forest archeologist in California, says she has seen Bigfoot on three separate occasions in the Ouachita Mountains of Oklahoma.

The first time was in 2012, when she and a group of three other people purportedly saw "a big one and a little one" completely covered in very dark hair walking towards them in broad daylight.

"I jumped up and I said: 'There they are!' and I ran at them," the 52-year-old from Sonora, Calif., told Global News.

"It startled them and they bolted up the hillside like nothing you've ever seen before. It was like they were on a bungee and somebody just let go.

"It was a little terrifying, in fact. They were so incredibly fast."

She said her group had a camera running but it only ended up filming her.

Accepted Science & Paradigms Which Are Likely Wrong

Where are Bigfoot Sightings?

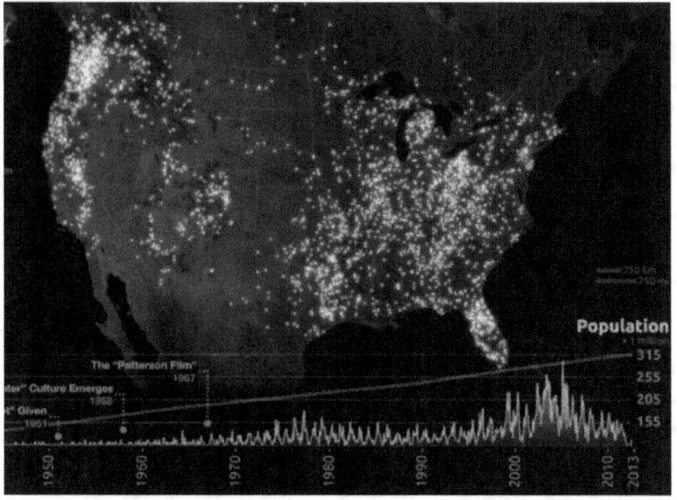

Reported sightings of Bigfoot — the legendary apelike creature that's been a favorite of cryptozoologists for decades — have abounded for decades. Now, for the first time, someone has created a map showing the places where alleged Bigfoot sightings have occurred.

Joshua Stevens, a doctoral candidate at Pennsylvania State University, used data compiled by the Bigfoot Field Researchers Organization (BFRO), which tries to document "the presence of an animal, probably a primate, that exists today in very low population densities," according to the group's website.

Stevens converted the BFRO data and, using geographic-information software, plotted 3,313 data points showing where people have claimed to see Bigfoot (aka Sasquatch, Skunk Ape, Yeti, Skookum or dozens of other names).

Accepted Science & Paradigms Which Are Likely Wrong

"Right away, you can see that sightings are not evenly distributed," Stevens said on his website. "There are distinct regions where sightings are incredibly common, despite a very sparse population. On the other hand, in some of the most densely populated areas, Sasquatch sightings are exceedingly rare. The terrain and habitat likely play a major role in the distribution of reports."

The map, which uses reports from 1921 to 2012, shows a plethora of supposed sightings in the Pacific Northwest, the Ohio River Valley, central Florida, the Sierra Nevada mountain range and the Mississippi River Valley.

Stevens' analysis also includes a chronological timeline showing a rise in reported sightings in the late 1970s (perhaps coinciding with the release of several B-movies about the mythical creature). Another spike in reported Bigfoot sightings occurred between 2000 and 2009.

Despite his exhaustive analysis of the BFRO data, Stevens stops short of giving the information more credibility than it deserves. "Ultimately, I'm not convinced there's a descendant of (giant ape) Gigantopithecus playing hide-and-seek in the Pacific Northwest," Stevens said. "But if respectable folks like ... primatologist Jane Goodall believe there's something more to the myth, I think it's at least worth putting on the map."

Accepted Science & Paradigms Which Are Likely Wrong

16.0 Genius is Only for a Few

I recently wrote a book about Genius titled "The Importance of Genius in our World".

My goal was to understand what Genius is, and why some people become recognized Geniuses and not others with high IQs.

I learned that there is a big difference between Genius potential and actual Geniuses because you have to accomplish something to be really recognized as a Genius.

Genius largely consists of the creativity and persistence to make a vision happen which creates a new creation, product, or process which can change the world.

It is also interesting to note that many scientists who study Genius think that individuals with I.Q.s from 125 and up can become accomplished Geniuses. This includes a lot of people who have the potential.

Accepted Science & Paradigms Which Are Likely Wrong

Can most individuals aspire to having Genius accomplishments or it only for the lucky few born with those qualities? Actually there are many ways for people to improve their creativity which can lead to Genius accomplishments.

I cite many creativity exercises in my book "The Importance of Creativity and How to Improve Yours". Here is just one of them:

A Creativity Exercise-Six thinking Head Coverings

This is another approach used to evaluate the optimization of a product or idea. In a group, an individual or small team "wears" one of the covers. When reviewing the idea in question, each "hat" maintains its assigned perspective:

Judgment: The judgment hat addresses the challenges or problems with the product or idea by considering the opposite point of view.

Emotion: The emotion hat represents the feelings or perceptions associated with the project or idea.

Logic: The logic hat represents the facts related to the product or idea.

Optimism: The optimism hat represents the possibilities for the product or idea with no barriers.

Creativity: The creativity hat introduces new ideas or possibilities for the idea or product.

Management: The management hat oversees the discussion and makes sure the team represents all perspectives.

Accepted Science & Paradigms Which Are Likely Wrong

17.0 Little People Are Just a Myth

There are many legends of Little People existing all over the world. And some of these legends have been proven by recent archeological work.

In this chapter I present information on Flores Man and a couple of the legends of Little People from other parts of the world. This information is from my book "About the Little People: Fairies, Elves, Dwarfs, and Leprechauns".

17.1 Flores Man in Indonesia

The skeletons of Flores Man have been excavated and they were only three feet seven inches tall as fully formed adult beings.

This was a race of real intelligent beings, and maybe these little people also existed in other areas around the world.

Accepted Science & Paradigms Which Are Likely Wrong

Homo floresiensis ("Flores Man"; nicknamed "Hobbit") is a species of small archaic human that inhabited the island of Flores, Indonesia, until the arrival of modern humans about 50,000 years ago.

The remains of an individual who would have stood about 1.1 m (3 ft 7 in) in height were discovered in 2003 at Liang Bua on the island of Flores in Indonesia. Partial skeletons of at least nine individuals have been recovered, including one complete skull, referred to as "LB1". These remains have been the subject of intense research to determine whether they were diseased modern humans or a separate species; a 2017 study concludes by phylogenetic analysis that H. floresiensis is an early species of Homo, a sister species of Homo habilis.

This hominin was at first considered remarkable for its survival until relatively recent times, initially thought to be only 12,000 years ago. However, more extensive stratigraphic and chronological work has pushed the dating of the most recent evidence of its existence back to 50,000 years ago. The Homo floresiensis skeletal material is now dated from 60,000 to 100,000 years ago; stone tools recovered alongside the skeletal remains were from archaeological horizons ranging from 50,000 to 190,000 years ago. Some people claim to have seen Flores Man in modern times.

Does Flores Man Live Today?

An ancient legend from the Indonesian island of Flores speaks of a mysterious, wild grandmother of the forest who eats everything: the 'ebu gogo'.

Accepted Science & Paradigms Which Are Likely Wrong

The 2004 announcement of a new branch on the human evolutionary tree was astonishing, to say the least. Standing just over a meter tall, the hominin labelled Homo floresiensis had a small brain, the apparent ability to make arduous water crossings, and seemingly honed skills in making stone tools. Much of the species' anatomy looked primitive, yet evidence for their behavior indicated an advanced, humanlike being. The hominin was so seemingly mythical that the research team drew from J R R Tolkien's fictional world for its nickname: the hobbit.

Arguably the strangest aspect of the diminutive hominins' story was the suggestion that they survived into the recent past, roaming the tropical forests and ancient volcanoes as recently as 12,000 years ago. Not only was this date surprising because it is a time when scientists believed that Homo sapiens were alone on the planet, but also because it was long after the arrival of modern humans in the area – tens of thousands of years after, in fact. Had hobbits lived alongside our own species for all that time?

Associations between ebu gogo and H floresiensis arose immediately after the media hobbit frenzy broke. From news headlines to scientific meetings, people wondered: could these two creatures be one and the same? Had the locals been imagining mythical, wild people of the forest – or merely reporting on them? Perhaps the seemingly fictitious legend had an empirical basis all along. While the media ran with the idea, some scientists, too, entertained it – fueling hope that the legend could suggest that a living, breathing H floresiensis could still be found on some remote part of the island today.

The proposed connection between the bones and the myth raised an interesting question, one that is being explored by anthropologists in other parts of the world: how far back in time can oral traditions accurately report events? Some scientists studying indigenous memory have suggested that oral traditions contain extraordinarily reliable records of real events occurring thousands of years ago. Where, then, are the boundaries between legend, memory, myth and science? Had the people of Flores preserved an oral record of H floresiensis?

The ethnographer who originally documented the tale of ebu gogo, Gregory Forth of the University of Alberta in Canada, argued that anthropologists are too inclined to dismiss folk categories as products of the imagination, while others pointed to the many correlations that existed between the description of ebu gogo and H floresiensis. Both were described as having long arms, for example, and being small in stature. Many were intrigued by the extreme detail of the legend; surely the vivid description of the 'pendulous breasts' that the ebu gogo allegedly threw over her shoulders must be compelling. Forth even lamented that the 'dimensions of female breasts is, unfortunately, one of many things that cannot be gauged from paleontological evidence'.

From the beginning, there were, however, weak links in the proposed connection between the prehistoric bones and the mythical legend. To begin with, the two concepts exist in entirely different regions of Flores. The category 'ebu gogo' belongs to the Nage people who reside more than 100 kilometres away from the H floresiensis discovery site at Liang Bua, across treacherous mountains and thick jungle forests. The hobbit cave is instead home to the culturally and linguistically distinct people known as the Manggarai. While it is not unimaginable that H floresiensis could have roamed the landscape, it is suspicious that ebu gogo is not a Manggarai invention. A quick glance across the archipelago also reveals that stories of small forest creatures are not unique to Flores, which is perhaps unsurprising given that the area is rife with living, humanlike primates. The well-known orang pendek (short people) of nearby Sumatra, for example, are thought to be accounts of orangutans. While Flores has no orangutans, there are plenty of macaques.

The myth persisted even as real scientists scoffed. But eventually holes in the ebu gogo/H floresiensis association grew too large to be ignored. Each expedition in search of a reported sighting revealed an empty cave or else, a macaque. New pieces of scientific evidence have also made the connection increasingly implausible, especially a revision of the dating that moved the hobbits' disappearance to almost 50,000 years ago. To experts, ebu gogo was about as real as the tooth fairy.

So, what then, are we to make of the legend of ebu gogo? Why are we so captivated by the idea of ancient wildmen of the forest?

Maybe the significance of the intertwined stories of H floresiensis and ebu gogo, then, is the realization that scientific discoveries – particularly the

Accepted Science & Paradigms Which Are Likely Wrong

unexpected ones – have the power to transform the way we think. By confronting scientists with something so unforeseen, these small bones opened the door to big speculation.

H floresiensis revealed that the past was more bizarre than we imagined, full of evolutionary hodgepodges, unexpected migrations, and life in surprising places. And while the legend of ebu gogo failed to echo paleoanthropological reality, such botched connections are not always the case. Researchers from geology to palaeontology turn to folklore, and events from volcanic eruptions to fossil discoveries have shown that science has something to gain from engaging with legend. Even the fabled creature with a lion's body and an eagle's beak introduced to Greek travelers as the griffin was likely grounded in encounters with dinosaur bones. The interplay between science and myth has become ever more complex – and more interesting. After all, if hobbits once lived on a remote Indonesian island, what else was once possible?

17.2 Other Little People Legends

Legends of Little People are worldwide. Here are just two of those examples:

<u>Iceland</u>

Huldufólk or hidden people are elves in Icelandic and Faroese folklore. They are supernatural beings that live in nature. They look and behave similarly to humans, but live in a parallel world. They can make themselves visible at will.

In Faroese folk tales, hidden people are said to be "large in build, their clothes are all grey, and their hair black. Their dwellings are in mounds, and they are also called Elves."

Accepted Science & Paradigms Which Are Likely Wrong

Some Icelandic folk tales caution against throwing stones, as it may hit the hidden people.

The term huldufólk was taken as a synonym of álfar (elves) in 19th-century Icelandic folklore. Jón Árnason found that the terms are synonymous, except álfar is a pejorative term. Konrad von Maurer contends that huldufólk originates as a euphemism to avoid calling the álfar by their real name.

There is, however, some evidence that the two terms have come to be taken as referring to two distinct sets of supernatural beings in contemporary Iceland. Katrin Sontag found that some people do not differentiate elves from hidden people, while others do. A 2006 survey found that "54% of respondents did not distinguish between elves and hidden people, 20% did and 26% said they were not sure."

Modern Accounts:

Do elves really exist? You bet they do, here in Iceland Elves have been a part of the folklore in Iceland since time immemorial, and if you were to ask any local, they will tell you earnestly that elves appear regularly to those who know how to see them.

Construction sites have been moved so as not to disturb the elves, and fishermen have refused to put out to sea because of their warnings: here in Iceland, these creatures are a part of everyday life.

But honestly, do they really exist?

Accepted Science & Paradigms Which Are Likely Wrong

Anthropologist Magnus Skarphedinsson has spent decades collecting witness accounts, and he's convinced the answer is yes.

He now passes on his knowledge to curious crowds as the headmaster of Reykjavik's Elf School.

"There is no doubt that they exist!" exclaims the stout 60-year-old as he addresses his "students", for the most part tourists fascinated by Icelanders' belief in elves.

What exactly is an elf? A well-intentioned being, smaller than a person, who lives outdoors and normally does not talk. They are not to be confused with Iceland's "hidden people", who resemble humans and almost all of whom speak Icelandic.

To convince sceptics that this is not just a myth, Skarphedinsson relays two "witness accounts", spinning the tales as an accomplished storyteller.

<u>An elvish warning</u>

The first tells of a woman who knew a fisherman who was able to see elves that would also go out to sea to fish.

One morning in February 1921, he noticed they were not heading out to sea and he tried to convince the other fishermen not to go out either. But the boss would not let them stay on shore.

That day, there was an unusually violent storm in the North Atlantic but the fishermen, who had heeded his warning and stayed closed to shore, all returned home safe and sound.

Accepted Science & Paradigms Which Are Likely Wrong

Seven years later, in June 1928, the elves again did not put out to sea which was confusing because there had never been a fierce storm at sea at that time of year. Forced to head out, they sailed waters that were calm but caught very few fish.

"The elves knew it," the anthropologist claims.

The other "witness" is a woman in her eighties, who in 2002 ran into a young teen who claimed to know her. Asking him where they had met, he gave her an address where she had lived 53 years ago where her daughter claimed she had played with an invisible boy.

"But Mum, it's Maggi!" exclaimed the daughter when her mother described the teen.

"He had aged five times slower than a human being," says Skarphedinsson.

Surveys suggest about half of Icelanders believe in elves.

"Most people say they heard (about them) from their grandparents when they were children," says Michael Herdon, a 29-year-old American tourist attending Elf School.

Iceland Magazine says ethnologists have noted it is rare for an Icelander to really truly believe in elves. But getting them to admit it is tricky.

"Most people tread lightly when entering into known elf territory," the English-language publication wrote in September.

That's also the case with construction projects.

Accepted Science & Paradigms Which Are Likely Wrong

Back in 1971, Skarphedinsson recalls how elves disrupted construction of a national highway from Reykjavik to the northeast. The project, he says, suffered repeated unusual technical difficulties because they didn't want a big boulder that served as their home to be moved to make way for the new road.

"They made an agreement in the end that the elves would leave the stone for a week, and they would move the stone 15 metres. This is probably the only country in the world whose government officially talked with elves," Skarphedinsson says.

But Iceland is not the only country that is home to elves, he says. It's just that Icelanders are more receptive to accounts of their existence.

"The real reason is that the Enlightenment came very late to Iceland.

<u>Fairies of Hawaii</u>

-The Menehune Fishpond

Accepted Science & Paradigms Which Are Likely Wrong

Menehune are a mythological race of dwarf people in Hawaiian tradition who are said to live in the deep forests and hidden valleys of the Hawaiian Islands, hidden and far away from human settlements.

The Menehune are described as superb craftspeople. They built temples (heiau), fishponds, roads, canoes, and houses. Some of these structures that Hawaiian folklore attributed to the Menehune still exist. They are said to have lived in Hawai'i before settlers arrived from Polynesia many centuries ago. Their favorite food is the mai'a (banana), and they also like fish. Legend has it that the Menehune will only appear during night hours, in order to build masterpieces. But if they fail to complete their work in the length of the night, they will leave it unoccupied. No one but their children and humans connected to them are able to see the Menehune.

The Menehune Fishpond, near Līhu'e, Hawai'i, on the island of Kaua'i, is a historic Hawaiian fishpond. Also known as Alakoko Fishpond, it has been listed on the U.S. National Register of Historic Places.

Also called Alekoko or Niumalu Pond, it is bounded by a wall 900 feet long at a large bend in Hulē'ia River. It has been deemed "the most significant fishpond on Kauai, both in Hawaiian legends and folklore and in the eyes [of] Kauai's people today. It is the largest fishpond on the island of Kaua'i. It is estimated to have been built in the 15th century and is believed to be the first brackish-water fishpond built in the Hawaiian Islands. It is so old that its construction is attributed to the Menehune, a mythical people inhabiting Hawai'i before the Hawaiians arrived....Additionally, it is the best example of an inland fishpond in the entire state." It was listed on the U.S. National Register in 1973; the listing included one contributing site and one contributing structure.

Accepted Science & Paradigms Which Are Likely Wrong

18.0 Memories Before Birth

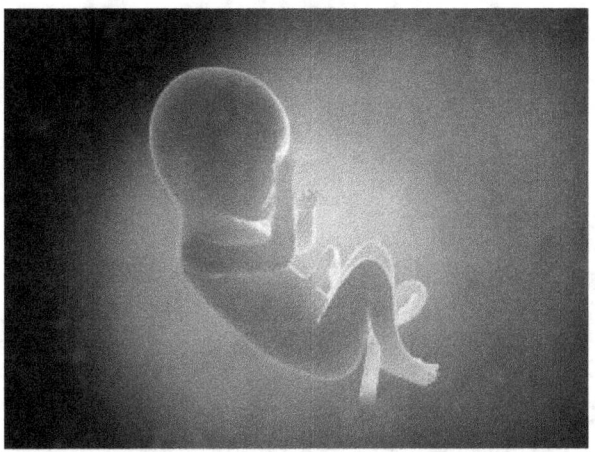

One of the big arguments in the United States today is if abortion is murder. Do babies have consciousness in the womb? Are they alive before birth?

If there is a way to determine the "life status" to help answer the abortion question don't we want to know the answer?

It is very difficult to prove that babies are really alive and conscious in the womb. There are numerous stories by people who have memories in the womb.

I remember being in my Mom's womb and would estimate that my consciousness was there for at least the second and third trimesters.

Here is my story as a form of subjective evidence:

Accepted Science & Paradigms Which Are Likely Wrong

What I Remember Before My Birth

Most people would start their life story when they were born because that is what they remember, or they think that is when their lives started.

But I remember from before my birth so we will start there….

My first thoughts in this life were that I was a disembodied consciousness outside and above the Earth. I seemed to be in space and could see the Earth. I enjoyed my state of being, but also felt I had a mission and reason to go back to Earth. One of my co-entities said that I had done enough on Earth and should not have to go there again. My spirit realized I had already spent many lifetimes on Earth. But I felt it was important for me to go back once more so I broke part of my consciousness off of my core and sent it towards Earth.

On Earth I was aware of my individual consciousness and was looking for parents to be born into their family. I found several prospective parents who were each ready for a new baby in upstate New York. I don't remember why that area was selected for the candidates and there were several couples to choose from. I would view each potential Mom and her husband using some type of ability looking forward. I could tell which couple would be the best one to raise me.

The third couple I viewed and looked at their timeline, and judged them to be the right parents for me to choose. Then I got close the Mother I had chosen and felt myself drawn into her womb. I remembered being inside her womb and growing. I could hear noises from the outside and movement.

Accepted Science & Paradigms Which Are Likely Wrong

After a timeless period my little world got smaller and smaller, which meant I was getting bigger and bigger. I could hear noises outside of my Mom as she went shopping or went to Church and other places. I became familiar with daily sounds.

Being Born and My Early Life

It was the year 1955 and my parents lived in the town of Painted Post in upstate New York in the United States. The United States was the greatest world power and my parents and their friends always mentioned how lucky we were to live in the U.S.

We lived in new house in the small industrial one company town. (Ingersoll Rand) Everything was picturesque and it seemed like it was right out of a Normal Rockwell painting.

I continued to grow in my mother's womb until I could feel the walls of it pressing in. Then I got tangled up in the baby's cord and was almost strangled by it. I managed to turn myself around so my head was up and feet down. This made it much easier for me to not get choked by the cord.

My room to move in the womb gradually seemed to get smaller until one day I kicked my feet and my mother's water broke. Then followed hours of contractions until I was forced out of my Mom's birth canal. It was extremely painful as the Doctor pulled me out by my feet. My head felt like it was being crushed in the tunnel and finally I popped out! (Being born is a very painful experience and uncomfortable. It is no surprise that most of us want to forget it.)

Accepted Science & Paradigms Which Are Likely Wrong

Though I was breathing on my own, the Doctor hit me on my bottom anyway and the pain caused me to take a huge breath which felt like breathing fire. I also started choking and the Doctor ended up putting me in an iron lung to help my breathing. After an hour or so the mucous in my throat and lungs were relieved and I could breathe normally.

A couple days later I was taken home by my parents and passed around for the grandparents to hold me. My father's mother didn't hold me right and I started crying because I was uncomfortable. Then my other grandmother who was a nurse made suggestions on how to hold me. I felt sorry for my paternal grandmother and so I stayed quiet when she held me again. She was happy because I was now peaceful in her arms.

Very quickly I also learned that when I needed something I would need to cry to get attention. I realized that my connection to the world was important and I needed to work on it to help me get food, my Mom's attention, and other things to live and be comfortable.

I later realized that the connection I was making with my parents to get their help and the rest of the world were the start of building my Ego. While a child and teenager I thought this self-centered Ego was who I was. In later years I learned this was not correct.

There were many more instances when I remembered my early childhood. A few months later my Dad was playing with me and tossing me into the air. Then Dad tossed me too hard and I hit the wall over the couch. I started crying but wasn't really hurt. I was just surprised and shocked by what happened. My Dad was very upset because he was afraid he had really hurt me. After that experience he was much more careful with me.

At age nine months I was playing in the front yard with my Mother. She was trying to get me to walk. She stood me up to get me on my feet and I started walking. The problem was I was going downhill and couldn't stop. I walked slowly, then was going faster and faster until I fell and started crying. But I had walked for the first time at nine months of age.

About age two I was watching cartoons on an early black and white television when I "Woke Up". It seemed like I had been in a real daze since I was born and now I was coming out of it. I looked around the living room and it was like I saw it all for the first time. Everything seemed more vibrant in my sight and I really felt alive for the first time. It seemed that I had been in some type of trance since my spirit came to Earth and now I had been released from the trance. From that point on I observed everything around me with pure clarity and asked the whys and wherefores of everything that happened to me.

At my age, happiness was being fed, feeling comfortable, and my mom holding me. I was still too young to understand all about what made one happy and sad living in the world.

Many years later I wondered why I was able to remember before my birth, the birth itself, and my early years. Most people say they can't remember those periods at all. My thinking is that I had some reason in living and being on the Earth. The question is what was that purpose?

Accepted Science & Paradigms Which Are Likely Wrong

Accepted Science & Paradigms Which Are Likely Wrong

19.0 Scientific Method and the Paranormal

I've had many paranormal Prophecy experiences in my life and have written several books on the topic. (See my book "Use Intuition and Prophecy to Improve Your Life-By An Adept" to learn more about my experiences.)

My first book was titled "On Using the Scientific Method to Study the Paranormal" which suggests a new paradigm for measuring paranormal phenomena.

Our scientific understanding of the world is only about 500 years old.

Mankind has existed for over 100,000 years, and the universe is billions of years old.

Is humanity so arrogant as to say that we have a close to final understanding of the natural scientific laws of the universe, or should we be more humble and admit that we only understand a tiny fraction of what is out there, and much more is undiscovered than discovered.

I spent many years reading articles and journals from organizations like the ASPR (American Society for Psychical Research) who have done good experimental work for 60 years on validating and understanding psychic phenomenon.

However if I were to go to the average person on the street they would say that these things have never been proven.

Accepted Science & Paradigms Which Are Likely Wrong

Most scientists would also say that paranormal events haven't been proven to exist.

Instead of exploring how to understand and benefit from these abilities, most researchers in these areas are still being asked to prove that these things really exist. In this case many of the skeptics aren't really interested in the objective evidence because it would disrupt their cozy worlds.

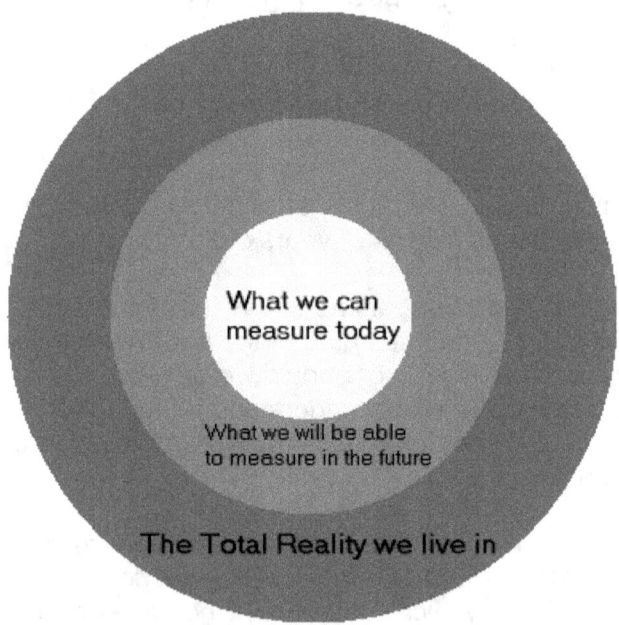

From the foregoing discussion on the scientific method and what is measurable, you can tell that I must have a significantly different idea of reality than the norm.

The diagram above best illustrates my belief of our ability to understand Reality:

Accepted Science & Paradigms Which Are Likely Wrong

The inner yellow circle represents what we can measure with our instruments today and perform experiments on to prove or disprove theories. The red circle is a larger area, which we will eventually be able to measure to understand and prove or disprove the way things are.

The blue outer circle is the largest area, and is that part of the universe which we may be able to experience but will never be able to measure and validate with objective scientific approaches.

We may be able to subjectively perceive a lot of things in the blue area, but will never have the tools and techniques to objectively quantify it.

This blue realm may also include such things as where the soul goes after death, the fundamental nature of God, and certain dimensions of space and time, which we can postulate but never prove or disprove.

The red region may be more amenable to creative approaches for objective measurement and validation.

However, there will have to be agreement among the scientific community on some new approaches, which may constitute legitimate standards for objective measurement of phenomena.

This may include indirect evidence, which is used in areas like particle physics.

Accepted Science & Paradigms Which Are Likely Wrong

Neutrinos for example can't be directly perceived, but their existence can be inferred by collisions with other particles, which make cloud tracks which we can directly perceive.

The same approach should be transferable to validation of something like telepathy, where the medium of thought transference may not be understood at this point, but it can be validated through well-controlled blind studies and statistics.

We may need to use individual experiences and statistics about those experiences for validation of experiments.

I think that this type of validation issue of the objectivity of an experiment also presents barrier to further scientific progress.

Accepted Science & Paradigms Which Are Likely Wrong

20.0 Summary

My goal in this book was to show as many concepts I question which are widely accepted but likely to be wrong.

Most people assume that if they read something about science in the news or a magazine that it must be true. This is not so. Scientists and Engineers are subject to personal bias and corrupt societal structures as many others are.

I've used my experience and the research I've done for my books over the years as the basis for my questioning many subjects here.

I hope you enjoyed the book and it raised lots of questions for you to think about.

All the Best,

Martin K. Ettington
August 2022

Accepted Science & Paradigms Which Are Likely Wrong

21.0 Bibliography

Ettington, M. K. (2001). *On Using the Scientific Method to Study the Paranormal.*

Ettington, M. K. (2008). *Physical Immortality: A History and How to Guide.*

Ettington, M. K. (2013). *The 10 Principles of Personal Longevity.*

Ettington, M. K. (2019). *Ancient and Prehistoric Civilizations.*

Ettington, M. K. (2020). *The History or Antidiluvian Giants.*

Ettington, M. K. (2021). *Bigfoot Mysteries and Some Answers.*

Ettington, M. K. (2021). *Four Evidences of Aliens and UFOs in Earth's History.*

Ettington, M. K. (2021). *Memories Before Birth and Reincarnation.*

Ettington, M. K. (2021). *Stranger Than Science Stories and Facts.*

Ettington, M. K. (2021). *The Importance of Creativity and How to Improve Yours.*

Ettington, M. K. (2022). *A 300 Million Year Old Civilization Existed on Earth.*

Ettington, M. K. (2022). *About the Little People: Fairies, Elves, Dwarfs, and Leprechauns.*

Ettington, M. K. (2022). *Planet Earth is Conscious.*

Ettington, M. K. (2022). *The Encyclopedia of Out of Place Artifacts.*

Ettington, M. K. (2022). *The Importance of Genius in our World.*

Ettington, M. K. (2022). *The Microscopic World Inside and Around Us.*

https://www.courthousenews.com/did-big-bang-really-happen-scientist-disputes-theory-of-universes-origin/#:~:text=According%20to%20Lerner%20%E2%80%93%20who%20wrote,for%20some%20time%20among%20astronomers. (2020). Retrieved from The Big Bang Never Happened.

https://www.discovery.org/a/24041/. (2015). Retrieved from The Top Ten Scientific Problems with Biological and Chemical Evolution.

https://www.forbes.com/sites/andreamorris/2022/05/05/5-things-science-is-getting-wrong-according-to-scientists/?sh=ee14022554aa. (2022). Retrieved from 5 Things That Science is Getting Wrong .

https://www.indiatimes.com/technology/news/humans-mars-ad-astra-nuclear-rocket-speed-550000.html#:~:text=Highlights&text=Planning%20to%20move%20to%20Mars,kilometres%20(300%20million%20miles). (2021). Retrieved from Taking a Nuclear Rocket to Mars.

https://www.lppfusion.com/science/cosmic-connection/plasma-cosmology/the-growing-case-against-the-big-bang/. (2022). Retrieved from The Growing Case Against the Big Bang.

https://www.nbcnews.com/science/space/maybe-dark-matter-doesn-t-exist-after-all-new-research-n1252995. (2021). Retrieved from Maybe Dark Matter Doesn't Exist Afterall.

www.ingramcontent.com/pod-product-compliance
Lightning Source LLC
Chambersburg PA
CBHW052357220526
45465CB00003BB/1133